TRASH CONFLICTS

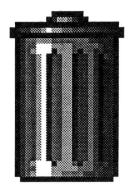

A Science and Social Studies Curriculum on the Ethics of Disposal

An Interdisciplinary Curriculum

by Amy Ballin
with Jeffrey Benson and Lucile Burt

esr

EDUCATORS
for
SOCIAL
RESPONSIBILITY

Educators for Social Responsibility
23 Garden Street
Cambridge, MA 02138
(617)492-1764

© 1993 by Boston Area Educators for Social Responsibility and Amy Ballin.
All rights reserved.
Published 1993.
Printed in the United States of America.
97 96 95   5 4 3 2

ISBN 0-942349-06-7

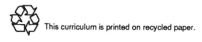 This curriculum is printed on recycled paper.

TRASH CONFLICTS

A SCIENCE AND SOCIAL STUDIES CURRICULUM ON THE ETHICS OF DISPOSAL

Table of Contents

Acknowledgments

Many people and institutions have been helpful in bringing this curriculum together.

Through the national center of Educators for Social Responsibility, The Streisand Foundation provided the funding for the first full draft of the curriculum. The Island Foundation supported the pilot program to test the material and support several local school systems' efforts to incorporate environmental education into their programs. The Toxics Use Reduction Institute supported the inclusion of material on reducing toxic waste.

Bill Kreidler's guidance and feedback in the initial stages of writing helped Lucile and me get on the right track.

Will Snyder introduced the issues of environmental inequality many years ago in our Boston Area Educators for Social Responsibility environmental study group. With Will's inspiration, Lucile and I began researching the issues in environmental justice that constitute Chapter 3.

Evelyn Lang of the Soloman Schecter School in Newton, Massachusetts gave me flexibility and support in developing the original form of this curriculum.

Francine Achbar of WBZ-TV in Boston has allowed us to make "Fire in Their Hearts" available to schools which adds inspiration to Activity 26.

Bradley Angel provided information and contacts on the hazardous waste incinerator controversy involving the Native Americans in Dilkon, Arizona.

Tim Greiner's expertise in hazardous wastes and toxic use reduction was invaluable. He introduced me to the Green Square Game and wrote the hazardous waste audit activity. Throughout the three years of work on this project, Tim has contributed to lesson titles, discussion questions and, most importantly, provided support and encouragement.

Lucile Burt's wonderful writing skills and creativity made this curriculum come alive. She contributed strongly to the background for teachers and discussion questions, making them comprehensive and thought provoking. Her sense of humor made working with her a true pleasure.

Jeffrey Benson, who joined this project mid-way through has provided insightful comments on several sections. He rewrote numerous articles and made them age appropriate and interesting for students.

Nancy Caldwell's ability to find mistakes and inconsistencies is amazing. Her precision in editing and clarity in pedagogy has been indispensable.

Several staff members and interns helped in this endeavor. Rachel Wyatt worked hours on formatting and editing the second version of the manuscript. She contributed many creative ideas that improved the curriculum. Bethany Chaney's research into environmental injustice in Southeastern states was insightful. Elizabeth Pritchard, Steven Benson and Sally Orme recruited the many permissions that were necessary. Erik Jacobson assisted with the glossary and bibliography. Ken Cherian prepared articles. Sharon Bassett's and Sally Orme's care and patience were crucial to complete the many tasks involved in the final stage of production, tasks ranging from reformatting to proofreading and spanning the table of contents through the glossary.

Finally, I would like to acknowledge Rachel Poliner, Chapter Director of Boston Area Educators for Social Responsibility. Rachel is responsible for making the curriculum possible. She secured the funding to pay for the writing, printing and teacher-training workshops. She has worked countless hours, late at night and weekends, editing and rewriting the entire curriculum. She has made contributions to every lesson. In particular, she brought together the section on Taking Action (Chapter 4). I can not thank Rachel enough for her writing, editing and commitment to the project.

Amy Ballin
February 1993

TRASH CONFLICTS

A SCIENCE AND SOCIAL STUDIES CURRICULUM ON THE ETHICS OF DISPOSAL

INTRODUCTION

Recently an eight-year-old girl told her mother that she didn't want to take juice boxes for lunch any more because they were bad for the environment. A high school senior, appalled by the amount of trash generated in the lunchroom, arranged for the recycling of styrofoam trays in the cafeteria. The growing awareness about environmental problems in our society is a concern young people share and want to do something about.

We believe strongly that education is the first step in making changes in habits and practices that are currently threatening the environment. Across the country, evidence is abundant that people are willing to change their habits once they understand the consequences of their choices. We see greater numbers of communities instituting recycling programs. Paper product manufacturers are beginning to advertise on national television their use of recycled paper in making their products. Consumers are a powerful group, and clearly, manufacturers are responding to an increasingly concerned, educated, and vocal public.

If young people grow up with both the ethic of preserving or recreating a safe environment and the knowledge of how to do that, they will carry that ethic and knowledge into their actions as adults.

Who This Book Is Designed For

This curriculum is for middle school students learning science and social studies. Math and language skills are also incorporated into many of the activities. The design of the curriculum allows teachers in traditional as well as non-traditional school structures to choose activities that will be appropriate.

The inclusion of materials that might ordinarily be taught as separate disciplines reflects our belief that an interdisciplinary approach to teaching about contemporary issues gives students a richer, more complex view. In the study of trash disposal, science students will benefit from understanding the social issues related to the problem. Social studies students need some scientific background in order to understand the social issues.

How This Book Is Organized

The curriculum is organized so that the first activities give a general sense of the scope of today's trash disposal problem. Students learn about the amount of trash that is created and the methods by which we dispose of it. Activities that might be classified as more "science" include experiments in which disposal methods are tried in the classroom. Activities that might be classified as more "social studies" include readings, discussions, and role plays about the controversies concerning various methods of disposal and use of resources.

Later activities focus specifically on hazardous waste and the issues of environmental safety and social justice related to its disposal. In several places we have included information about actions that have been taken by individuals, groups, and industries to address the problem of hazardous waste disposal. We also suggest some actions that students themselves can take and guidelines for getting involved in such projects.

How To Use This Book

We have designed these materials so that teachers may choose individual activities from this book to include in other science or social studies curricula, or they may use the material here as a central focus for an entire semester course on scientific and social aspects of trash disposal. Depending on the needs of a given class, certain activities and articles may be skipped entirely or presented in a different order.

We encourage teachers to use other materials in conjunction with *Trash Conflicts*. With the issue of trash disposal as a context, students may be more interested than ever in other science topics, such as geology, ecology, and biology. The same might be true of social studies curricula. For example, if your state has considered enacting a Bottle Bill, whether it has passed or not, this would be an interesting case study for the civics topic of how proposals are (or are not) enacted into law.

At the beginning of each activity, the **Objective** states what students will do and should learn. In addition, we sometimes suggest activities that students ought to do prior to the lesson so they will understand the material. Each lesson includes the **Time** required, **Vocabulary** used, **Materials** needed, and the **Activity Procedure**. An activity may also include a **Note,** which alerts teachers to special concerns or requirements, and **Supplementary Information for Teachers**, which provides facts and explanations relating to the activity's focus. Student handouts are identified as such on the first line of each handout page and are in Helvetica font (this is Helvetica). A **Glossary** at the end of the book contains all the vocabulary words listed in the activities.

Assessment and Evaluation

There are numerous opportunities in *Trash Conflicts* for teachers to implement creative assessment tools. Throughout the curriculum students will develop products, give oral presentations, write reports, etc. Students can keep copies of letters they write and send to companies or government agencies, posters they create and display in school, samples of the recycled paper they make, and the audits and recommendations they fill out. All of these items can be compiled into student portfolios.

The "Taking Action" project at the end of the curriculum can be seen as a performance assessment tool. Teachers will see how much information students have absorbed and can apply, and whether they have learned where more information and other resources can be found.

We encourage teachers to have students keep journals. This is an excellent self-assessment tool for young people, and it will help teachers know what pressing questions or concerns students have.

Evaluation may be done by grading worksheets or projects involved in this curriculum, designing tests to determine whether students can apply certain aspects of the material, or asking students to write summaries about what they are learning.

Thinking Ahead About Students' Reactions

The study of waste disposal, like so many other contemporary issues, can make students feel powerless and despairing because there appears to be so little they can do. The world seems complex and offers them few opportunities to effect significant change. Understanding that people are working to solve the problem and that students can become part of the solution enables them to cope with feelings of helplessness.

When students practice thinking skills and decision-making about controversial issues, they are preparing to become active, participating adults. The lessons in this book give them practice at forming opinions and speaking out about what they are learning.

While we believe that being informed is important and that student action is empowering, we also believe that young people should not feel responsible for the decisions that adults make. When students learn about the problems and controversies concerning trash, as well as other complex environmental issues, they may see the information as evidence of another failure on the part of adults to take care of them and their world. It is important that they be aware that concerned adults are doing something.

At some point, it might be useful to discuss with students what their roles might be as young people who are educated about trash disposal. They need to be reassured that gaining knowledge and trying to make changes in their own developing habits of consumption and waste production are their most important jobs now. However, many adults are only now becoming informed about the very problems that young people are learning about, so students can be part of the dialogue in their families and communities on how to address the waste disposal problem.

It is possible that certain students may feel shame or be stigmatized by other students as they become aware of their own habits regarding trash disposal. They may also be reluctant to share information concerning where they live and what kind of work members of their families do; for instance, a student may live near a controversial toxic treatment facility or have a parent who works in one. Teachers may choose to speak privately with certain students before beginning an activity, or work with the entire class on accepting differences in personal and family backgrounds. Lessons may also be adapted and edited.

Including the Family and Community

Many of us are just beginning to face the trash problem as a nation and a world. Adults are often as uninformed as students may be when they begin this study. In the right atmosphere, adults and young people can learn together.

We recommend including the families and the wider community in this learning experience. As students explore how waste is generated, they will scrutinize their own and their families' habits. It is therefore helpful for the families of students to be involved. If families are included in the learning and examining of habits in a manner that does not leave them feeling blamed for any wasteful practices, then conversations between students and family members will be mutually enlightening and constructive. Parents can be kept informed of class activities by letters explaining assignments and encouraging them to participate in the projects their children do at home. They might also be invited to see class projects or to hear speakers. Students can keep the community informed about their projects and learning experiences through articles or letters to the local paper. Here too diversity in family backgrounds and habits should be anticipated, and time should be spent with students on learning to accept their differences.

Teachers are encouraged to incorporate field trips into this curriculum. Visiting a waste-to-energy facility, a local landfill, or recycling center can make many of these issues more real to the students.

We have included two letters below as models that teachers may use. One is from a teacher, informing parents of the activities students will be assigned; the other is from the Massachusetts Environmental Affairs Office, explaining to parents the importance of education about trash disposal. We believe that most parents will support and encourage this study when they are included as much as possible. Involving parents and community in this and other curricula also helps emphasize that education is not something that happens only in schools but is part of the life of families and the entire community.

To: Grade Six Parents
From: Amy Ballin, Science Teacher
RE: Upcoming Homework Assignments

The sixth grade is currently studying solid waste management in science class. To learn about this topic effectively, students need to understand waste management as it pertains to their lives. For this reason, some homework assignments involve looking at their own trash and the trash from their homes and town. The following homework assignments are samples of the kinds of activities included in this unit.

* Keeping a record of trash thrown out at home and in school for one week.

* Finding out where the trash goes in their town.

* Surveying shoppers in a supermarket to consider consumer habits.

* Conducting a household hazardous waste audit to locate hazardous chemicals in the home.

I realize some of these assignments may cause some inconvenience for your family. I will tell the children to do the assignment in the ways that work best for them and you. Your support and encouragement on these assignments is greatly appreciated.

On the back of this memo is a letter from Willa Small Kuh, Acting Director of the Massachusetts Department of Environmental Protection. Solid waste management is a new problem that needs to be addressed in school. I hope this letter will explain the importance of educating the next generation about this crucial topic.

Please call me with any questions you may have. I will let you know when we have speakers coming who might be of interest to parents.

Thank you very much.

DANIEL S. GREENBAUM
Commissioner

The Commonwealth of Massachusetts
Executive Office of Environmental Affairs
Department of Environmental Quality Engineering

Division of
Solid Waste
Management

1 Winter Street
Boston, MA 02108

L. James Miller
Director

Dear Parents:

For most Massachusetts residents, garbage disposal has been out of sight and out of mind. In urban areas, trash left at the curbside in the morning disappears, as if by magic, by evening. While the actual process of garbage disposal may still be out of sight, its burden on municipal budgets and the public health no longer allows it to be out of mind. Citizens are becoming more aware of the issues regarding solid waste management through frequent media coverage about topics such as recycling, the use of disposal diapers, and excess packaging.

The Department of Environmental Protection (DEP) believes that it is essential to teach young people why we should reduce the amount of waste we make and how it should be managed. The Department has a commitment to assist school education programs regarding solid waste management. It is exemplified by DEP's "Solid Waste Management School Resource Guide" for teachers which will be released in January 1990.

The Department commends educators who are committed to teaching this issue and those parents who support these efforts. I encourage the Solomon Schecter School to continue its excellent program of teaching solid waste management and to raise student awareness on all environmental issues.

Sincerely,

Willa Small Kuh

Willa Small Kuh
Acting Director

CHAPTER ONE

TRASH AND ITS DISPOSAL

OVERVIEW AND BACKGROUND INFORMATION FOR TEACHERS

OVERVIEW

The activities in this first chapter are designed to move students through several stages:

A. Realizing how much trash we as individuals and as a society produce, and that the amount of trash we produce is a problem.
>   * Activity 1 has students collect, sort, and measure their families' trash for a week.
>   * Activity 4 helps students learn about both the environmental and the economic problems surrounding our use of resources.
>   * Activity 9 includes a brief history of some trash disposal problems.
>   * Activity 11 has students read about the growing problem of trash disposal.
>   * Activity 12 includes the market problems around recycled paper.

B. Reflecting on our habits as consumers of resources and as producers of trash.
>   * Activity 1 has students compare their own trash habits with national averages.
>   * Activity 2 involves interviews of older adults or people from other cultures to hear about their former consumer and trash disposal habits.
>   * Activity 3 helps students learn about renewable and non-renewable resources and consider their own use of resources.
>   * Activity 10 has students find out where their own trash is taken when it leaves their homes.
>   * Activity 13 asks students to consider changing their own daily lunch habits.
>   * Activity 14 involves surveys of shoppers about their use of paper versus plastic bags.

C. Learning about and experimenting with the disposal and reduction of trash.
>   * Activity 5 has students create mini-landfills and analyze the results one month later.
>   * Activity 6 has students create a compost pile and analyze the results six months later.
>   * Activity 7 has students learn about leachate involved with landfills through a water cycle experiment.
>   * Activity 8 includes observing the burning of a small amount of trash.
>   * Activity 12 has students make recycled paper.
>   * Activity 13 challenges students to try to reduce the amount of trash from their own lunches.
>   * Activity 14 involves a comparison of the wastes involved with both the production and the use of plastic and paper bags.
>   * Activity 15 has students consider packaging, how to reduce it, and how to write to companies with suggestions.

BACKGROUND INFORMATION ON DISPOSAL METHODS

In order to understand some of the issues surrounding the disposal of waste in the United States, it is necessary to know about the current options for disposal. The Environmental Protection Agency (EPA) recommends that communities use what is called an "integrated waste management system": a combination of source reduction, recycling, combustion, and landfilling. The idea is to reduce the amount and toxicity of waste by recycling and source reduction, thereby reducing what must be disposed of in landfills and incinerators.

In 1988 the United States landfilled 80% of its waste, incinerated 9%, and recycled 11%. At that time the EPA. set goals for 1992 of landfilling 55%, incinerating 20% and recycling 25% of U.S. waste. The most recent figures available, from 1990, show that the U.S. landfilled 66.6%, incinerated 16.3%, and recycled 17.1% , showing some progress toward the goals set. The goals have not been without controversy, however, as many environmental organizations oppose greater reliance on incineration because they are concerned about pollution-causing emissions and the ash byproduct, which must be landfilled. This is explained further under "Incineration" below.

Two Ways of Reducing Amounts of Waste

1. Source reduction

Source reduction means lowering the amount of toxic materials and the volume of trash *before* it enters the waste stream. For the manufacturer, making such reductions involves using non-toxic substitutes where they are available, minimizing packaging of products, and recycling some by-products via other manufacturers. For consumers, source reduction means purchasing long-lasting rather than disposable products (pens, razors, etc.) and products with the least packaging. A combination of education and incentives will be required to make substantial changes in source reduction.

2. Recycling

Recycling is the collection and processing of materials for reuse either in the same form or as part of a different product. Recycling is experiencing an increase in popularity due to growing public awareness of the waste problem and municipalities reaching the limit of their disposal options. Currently it is possible to recycle paper, glass, aluminum, some plastics, steel, oil, and tires. Yard and food waste can be recycled through composting. Many communities have recycling centers where materials may be dropped off, and more are organizing curbside pickup service for recyclables.

One problem that has existed is the gap between the supply of recyclable materials and the demand for recycled products, as determined by consumers and manufacturers. *Newsweek* reported the problem in this way: "Thousands of residents (in Minneapolis) eagerly turned in their glass, cans, and newspapers. But newsprint handlers were so inundated that rather than buying it at $12 a ton, some started charging $20 a ton to haul it away. Only eight U.S. paper mills are equipped to turn old newspapers into new newsprint..." (*Newsweek*, 11/27/89, p. 69). Although markets and manufacturers of recycled products are growing, it is important to remember that any materials that can't be reused will eventually end up in landfills or incinerators.

Two Methods of Waste Disposal

Even when the amount of waste is reduced through source reduction and recycling, there will still be waste to dispose of. The two options are landfilling and incinerating (newer processes involving burning are called waste- or trash-to-energy plants), and currently there is heated debate about which is the better means of disposal. Each has advantages and disadvantages.

1. Landfills

Landfills are what used to be called the town dump. Waste is trucked to the site and dumped there. In recent years, federal and state governments have created new requirements for the design and operation of landfills so that the risk to the environment, and to humans who are part of the environment, is minimized. Many states require (1) lining the landfill with clay or plastic to prevent toxic liquids from entering the water supply, (2) covering each day's layer of waste with sand, and (3) covering the top with a thick layer of soil and landscaping.when the landfill is closed.

There are three major problems with landfills: lack of space, groundwater contamination, and methane production.

The most obvious problem is that many areas, particularly those which are densely populated, simply don't have enough space to landfill waste. Old dumps and landfills are filling up, and land for more is unavailable. Sometimes a town will pay to have waste trucked to a landfill that is quite far away. This clearly adds to the cost of waste disposal.

Ground water contamination occurs when precipitation falls on the landfill and runs into the ground beneath, carrying toxic materials with it. This contaminated water is called leachate. Depending on the geological makeup of the area, contaminated water can travel underground and enter the aquifer and the drinking water supply. In modern lined landfills, leachate is collected and pumped to a waste water treatment plant.

When waste decomposes in landfills, methane gas is produced. Methane, which is flammable, can cause fires or explosions if it is trapped, or can move laterally and pose risks to surrounding neighborhoods. In some modern landfills, methane is collected and sold to gas utility companies. (Methane is the largest component of natural gas.)

2. Incinerators and Trash-to-Energy Facilities

Formerly, incinerators were facilities that simply burned unsorted waste. Some such incinerators still exist. More modern incineration facilities use processed waste that allows more efficient and uniform burning. Both the old and new types of incinerators may be trash-to-energy or resource recovery facilities. This means that energy is generated from burning the waste.

Incineration has certain advantages, including an answer to the space problem created by landfilling. Incinerators are also less likely to attract vermin and create unpleasant odors. The creation of electricity as a byproduct can be useful to many communities.

Yet incineration is not without problems or controversy. Most people concerned about the growing use of incineration as a disposal method are worried about the emissions and the ash. Emissions are the substances that enter the air with the steam released from the burning. Ash is the residue that remains after burning and may contain concentrated amounts of toxic substances. Such ash needs to be disposed of, and often it is landfilled.

People who support incineration say that the toxicity of emissions and ash are within safe standards. But opponents maintain that not all dangerous substances are regulated and that many trash-to-energy facilities don't adhere to strict enough standards on the contents of emissions and ash.

One controversy has to do with whether the ash byproduct is hazardous. The EPA does not classify it as hazardous. There are two kinds of ash byproducts involved -- bottom ash and fly ash. Bottom ash, which is not labelled as hazardous, is the ash left after trash is burned and metals are separated out. The hazardous constituents in bottom ash, like lead, cadmium, and chromium, are bound up in a matrix of non-hazardous materials, much the way sand is bound up in concrete. The EPA test doesn't detect toxicity in these bound materials. Fly ash is the lighter ash that goes up the smoke stack, is caught on pollution control equipment, and then must be removed from the equipment. The fly ash is labeled as hazardous because the hazardous constituents are not bound to any other materials. When fly ash is disposed of in combination with the bottom ash, it is not labeled as hazardous.

Another controversy involved with incinerators and waste-to-energy plants is that they require huge amounts of trash to reach and sustain the necessary temperature for incinerating. Some people believe that this need for trash undermines attempts to change people's habits toward reducing and recycling. In fact, when large waste-to energy plants were built in the last decade, some towns signed contracts with the plants promising certain amounts of trash. When those towns started to implement recycling programs and sent less trash to the waste-to-energy plants, the plants fined them large penalties for not sending their promised amount of trash. This "put or pay" clause is not legal in Massachusetts any longer, but many towns that signed contracts during the 1980's are now paying penalties for sending less trash to waste-to-energy plants. Teachers and students may want to investigate if this has been a problem in their states.

Currently the issue over whether incineration is an answer to waste disposal problems is hotly debated. Supporters believe that we have enough information that proves incineration to be safe, and our trash disposal problem is so acute that we need to go with the best technology available at the moment. Opponents believe that we know too little about the long-term potential health threats to use incineration widely.

While debate continues, we continue to generate trash that we must dispose of. In the United States, we have developed throw-away habits that have vastly increased our amount of trash. At the same time, our increasingly urban population has little contact with its waste after it is set out for collection. We have thrown it "away," and many of us have no idea where "away" is. While it is important to develop safe, efficient methods of disposal, it is clear that we need to change our habits. Being informed about ways to reduce and recycle waste, making decisions as consumers to purchase lasting, reusable products with little packaging, and understanding the issues related to trash disposal methods are now important aspects of our responsibilities as citizens.

Activity 1

HOW MUCH TRASH DO YOU THROW AWAY IN A WEEK?

Objective

Students will become more aware of the content and amount of trash generated in daily life by observing and recording how much trash their families produce and exactly what is thrown away.

Note

It may be difficult to obtain some families' cooperation for students to collect trash from their homes because of embarrassment about the amount and kind of trash, issues of privacy, or inconvenience. The sample letter included in the introduction may help in soliciting parent cooperation with this project. If it is not acceptable for students to keep track of the family trash for a week, they may keep a record of their personal trash for the week without including the rest of the family, or simply keep a record of what they throw away in school. Students are asked to weigh the trash in order to compare their findings with the national statistics, which are recorded in pounds and tons (not in volume or metric weight). Care should be taken that individual students are not stigmatized by the amount and types of trash they are throwing away.

Time

20 minutes to introduce the procedure
1 week homework assignment (waste record)
45 minutes for follow-up activity. (see Activity Procedures - Part B)

Vocabulary

garbage   renewable resource
non-renewable resource   trash

Materials

1 copy of Handout 1, "Waste Record," for each student
scale (in pounds)
boxes or bags for sorted trash
pencils
calculators
1 copy of Handout 2, "Analyzing Waste Record," for each student (includes
        graphs, "Finding a Place for Trash" and "One Person's Trash in a Day...")

Activity Procedure - Part A

1. To get students to begin thinking about trash and where it goes, discuss the following questions:

> *What is trash?
> *What types of trash have you thrown away today?
> *Can you develop trash categories?
> *What would you guess is the category that generates the most trash?
> *Where did the trash go after you threw it away?
> *What happens to it at "away"?

2. Explain to students that, beginning the next day, they will keep track of everything they and their family throw away for one week. This is a way for students to begin to (1) understand how the issue of waste disposal has become a large problem and (2) to see their role in the trash crisis.

3. Distribute Handout 1, "Waste Record." (Students may need multiple copies.) Explain that on the handout they will list everything that is thrown away according to category, items, and weight in pounds. There are a number of methods they might use to keep track; students and their families should choose the best way for them. One method is to keep the waste record next to the trash and record each item as it is thrown away. In this case weights will have to be estimated, making the record much less accurate. Another way to keep track is to have boxes or bags for each category of trash and weigh each container at the end of each day or at the end of the week.

4. Before they begin, ask students to look at the record sheets and estimate which category they think will have the most trash by weight.

5. Students should bring their completed sheets to class at the end of the week.

Suggested activities during the week while students are keeping trash record:

> 2. Changing Times, Changing Waste
> 3. Identifying Natural Resources
> 4. Thinking About *The Lorax*

Activity Procedures - Part B

This activity should be done after students have had a week to collect data of their family's trash on Handout 1. Although the worksheet could be assigned for homework, we have discovered that some students have difficulty with the calculations and so find it easier to do the activity in class. It might be helpful to recruit a math teacher to assist students with these calculations

1. Have students work on Handout 2, "Analyzing Waste Record," individually or in pairs. If some students were only able to keep track of their own trash, they will find the calculations difficult and might benefit from using another student's data.

2. If students have not done Activity 3: Identifying Natural Resources, explain the concepts of renewable and non-renewable resources, or skip question 4 on the worksheet.

3. When students have finished work on Handout 2, conduct a general discussion in which students report on their findings. Include a consideration of how their analysis for one week may be different from the national averages. For example, yard wastes might be exceptionally high or low during a particular season. The national statistics given are based on a yearly average.

Name _____

WASTE RECORD

Food items	Pounds	Metal items	Pounds	Yard waste - list kind	Pounds
Food items recycled	**Pounds**	**Metal items recycled**	**Pounds**	**Yard waste recycled**	**Pounds**
Other items	**Pounds**	**Other items**	**Pounds**	**Other items recycled**	**Pounds**

Handout 1 Continued

Paper Items	Pounds	Plastic Items	Pounds	Glass Items	Pounds

Paper Items recycled	Pounds	Plastic Items recycled	Pounds	Glass Items recycled	Pounds

Name_____

ANALYZING WASTE RECORD

1. What three things were most frequently thrown away?

a.)_____

b)._____

c.)_____

2. What three things were least frequently thrown away?

a.)_____

b.)_____

c.)_____

3. What was recycled?

4. After each item on your list, indicate whether it is a renewable (R) or non-renewable (NR) resource.

5. Write the total number of pounds thrown away and recycled in each category:

	Thrown Away	Recycled
food	_____ pounds	_____ pounds
metal	_____ pounds	_____ pounds
yard waste	_____ pounds	_____ pounds
paper	_____ pounds	_____ pounds
plastic	_____ pounds	_____ pounds
glass	_____ pounds	_____ pounds
other	_____ pounds	_____ pounds
TOTAL	_____ pounds	_____ pounds

6. It is estimated that the average family of four in the U.S. produces 150 pounds of trash each week.

   150 lbs. x 52 weeks = 7,800 lbs. per year per family of four

This represents almost one ton of trash per person because

   7,800 lbs / 4 people = 1,950 lbs. per person (1 ton = 2,000 lbs.)

If you can't imagine what one ton weighs, think of how much 65 fourteen-inch TVs would weigh; that's about a ton!

   a.) How many pounds of trash does your family throw away each week? (do not include recycled items) _____

   b.) How much trash does your family throw away in a year? (total trash from line "a" x 52 weeks)_____

   c.) What is the average amount of trash each member of your family throws away in a year? (total from line "b" divided by the number of family members)_____

   d.) Does your family throw away more or less than the national average?_____

7. What is the *percent* of each category in the total trash? (For each category, divide the number of pounds by the total weight of the week's trash.)

	Thrown Away	Recycled
food	_____%	_____%
metal	_____%	_____%
yard waste	_____%	_____%
paper	_____%	_____%
plastic	_____%	_____%
glass	_____%	_____%
other	_____%	_____%

TOTAL = 100%

8. How did the amount of trash in each category compare with what you expected to find?

9. Below is a circle graph which shows the percentage of trash in each category thrown away by the average family. Make a circle graph for your family below, using the percentages from question 7.

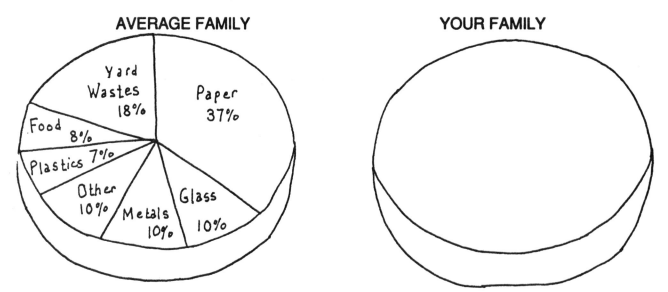

AVERAGE FAMILY

Yard Wastes 18%
Paper 37%
Food 8%
Plastics 7%
Other 10%
Metals 10%
Glass 10%

YOUR FAMILY

10. Compare your circle graph with those of other students. In what ways were the results similar? In what ways were they different? What was the biggest difference? Describe and explain any big differences (e.g., disposable diapers because there's a baby in the family, fewer yard wastes because you live in an urban area, a lot of yard waste because of the season, etc.)

11. Were the results in your graph similar to the national average? Explain why or why not.

12. Did you get the results you expected? Explain any differences.

13. Look at the graphs "Finding a Place for Trash" and "One Person's Trash in a Day."

     a.) Why do you think the amount of trash has changed?

     b.) Why do you think the use of energy recovery has increased? (Energy recovery is a process where energy is produced by burning solid waste. It is also called resource recovery.)

     c.) Use of landfills is not expected to increase in the year 2000. Why is this?

14.) What did you learn from this activity? Has your knowledge of the amount and content of trash changed? Explain.

15). Did you see ways to reduce the amount of trash your family threw out? For example, were there any things thrown out that could have been reused?

Finding a Place for Trash

Discards and recovery of houshold, commercial and industrial waste in millions of tons.

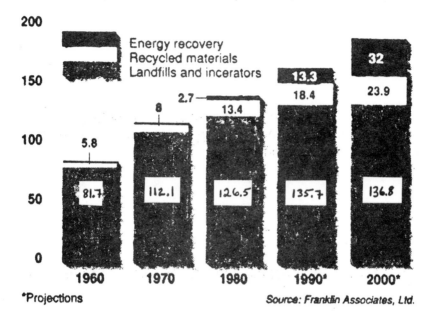

*Projections

Source: Franklin Associates, Ltd.

One Person's Trash In a Day . . .

Pounds of houshold, commercial and industrial waste generated per person in the United States per day.

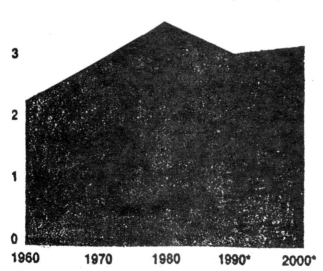

*Projections for waste generation without additional recylcing measures

Source: United State Environmental Protection Agency

Reprinted by permission of The New York Times

Activity 2

CHANGING TIMES, CHANGING WASTE

Objective

By conducting and reflecting on interviews of people from different generations, students will understand how changing values and technology have altered our consumption habits and therefore our waste production.

Suggested Previous Activity

1. How Much Trash Do You Throw Away in a Week?

Time

30 minutes to introduce assignment
1 hour homework assignment (interview)
20 minutes for follow up discussion

Vocabulary

trash

Materials

paper and pencils
interview questions (generated in class)
tape recorder (optional)

Activity Procedure

1. Explain to students that they will be interviewing adults to find out how the content and amount of trash has changed over time. In small groups, or as a whole class, have students brainstorm questions they think will help them find out about what people used to throw away. (Some sample questions are listed below.)

2. Help students identify a relative, neighbor, or family friend who might be interesting to interview. An individual who is several decades older than the student and/or who lived in another country or culture might have some very surprising responses.

3. To practice interviewing skills, have the class interview the teacher or other school staff with whom they feel comfortable.

4. Give students the following instructions:

Interview your parents, grandparents, or other older adults to find out what products they used in their everyday lives when they were your age and how those products were packaged twenty, thirty, forty, or fifty years ago. How were their lives different from yours? It would be interesting to interview any people you know who have lived in a country other than the United States.

When you are interviewing a person, it is important to make him or her comfortable. Plan ahead and choose a time that is convenient for the person you want to interview. Tell the person the subject of the interview and how long it will take (approximately 30 minutes).

If you have a tape recorder, you can tape the conversation. Be sure to ask permission of the person you're interviewing. Before our class discussion, write down some notes from your tape so that you remember clearly what the person said.

Sample questions for interviewing an older adult: (Encourage students to create their own questions. Use these as needed to get them started on various trains of thought.)

1. When you were my age, how did you dispose of trash? (Be sure to record the year being talked about.)
2. What kind of container did you use to hold trash?
3. How much trash did you generate each week?
4. How did you buy vegetables? meat? milk? How was each packaged?
5. Did you buy packaged pre-made meals (like TV dinners)?
6. What did you do with leftover food?
7. What did you do with old clothes?
8. How many cars, radios, televisions did your family have? How long was each used?
9. What kinds of toys did you have? What materials were they made of? What did you do if a toy broke?
10. What kinds of cleaning products did you use in the house?
11. What types of razors and pens were used? How long did they last?
12. How was the lawn mowed? What did you do with yard waste (dead leaves, grass clippings)?
13. If you were away from home and stopped for a soda or cup of coffee, how was it served? Where did you drink it? What did you do with the cup/bottle/glass when you finished?
14. Was anything recycled?

Suggestions for interviewing a person from another country: (In this situation, students should begin with these questions and then continue with the ones above.)

1. How do you think growing up in your country was different from growing up in the United States?
2. In what ways do you think garbage disposal was more or less of a problem in your country?

5. Ask students to reflect on the interviews, choose examples that were especially interesing to them, and summarize what they learned. As part of their homework, or in

class before discussion, have students write a paragraph about their overall impression of how consumption and disposal have changed. They can share these impressions and specific examples that stood out for them. Perhaps the class will want to choose one specific question that they all asked and hear the answer each student received.

6. Conduct a class discussion about the interviews. The following questions may be helpful.

* What conclusions can you draw about what was important in products and consumption in other times and places?
* How is your life different from that of the person you interviewed?
* What impact does this change have on how and what we buy and throw away?
* What factors do we consider in making our purchases? (Examples: cost, quality, convenience, brand name, appearance.)
* How important a factor is convenience in how we make our choices as consumers?
* In the time period described by the person you interviewed, were consumption and disposal habits less damaging to the environment than you think they are now? Were they any worse for the environment? Explain.
* Would you be willing to give up the convenience of disposable products? Think of examples of what you would be willing to give up and what you would not.

Activity 3

IDENTIFYING NATURAL RESOURCES

Objective

By tracing everyday objects back to their origins, students will discover that everything comes from the earth. In the process they will define what a natural resource is. Students will also distinguish between renewable and non-renewable resources and reflect on the implications of natural resource use.

Time

45 minutes

Vocabulary

natural resources   recyclable
non-recyclable   renewable resources
non-renewable resources

Materials

colored markers or pencils
dictionaries
encyclopedias
paper and pencils

Activity Procedure

1. Write the phrase "natural resources" on the board. As a class, come up with a definition and examples. This definition can be revised as the lesson continues. Write the list of examples on the board. (Examples: oil, iron, tin, cotton, trees, aluminum)

2. Once the list is on the board, ask students to identify which resources are renewable, meaning that we can replace them by growing or making more, and which are non-renewable, meaning that there is a limited supply which cannot be replaced. (Examples: oil is a non-renewable resource; a tree is a renewable resource.)

3. Have students make lists of what they are wearing and then try to trace each item back to its natural resource. (Examples: Leather shoes come from animal skin. Vinyl shoes are made from plastic, which is manufactured from petroleum, which is drilled from the earth.) Students can use dictionaries and encyclopedias to search for this information.

4. Once students understand the concept of natural resources, have the class return to the original definition and decide if they are satisfied. They should revise their definition as they see fit.

5. Have students make a list of things they used from the time they got up this morning until now. This includes getting up, showering, dressing, eating, and getting to school. It will help for them to remember all of their activities first and then figure out what they used for each one. (Example: Brushing teeth involves a toothbrush, toothpaste, water, and a glass).

6. When students have completed their lists, have them trace three of the items that they used back to their natural resources and, for each, decide if the product comes from a renewable or non-renewable resource. Students can help each other and use dictionaries and encyclopedias for this section.

7. When students have completed tracing their three items, they can each choose one to share with the class. (If there are certain items students can't trace, these might be done for homework -- see step 9 below.)

8. Hold a class discussion to take a final look at the definition of natural resources. Consider these questions :

    * Are you satisfied with the definition? If not, how would you change it? (Help students at this point with the glossary definition if necessary.)
    * Can you think of any item that can't be traced back to the earth? (There aren't any).
    * What problems could arise from the use of non-renewable resources?
    * Choose one renewable and one non-renewable resource (such as trees and oil); what would happen if these resources were used up?
    * What are some ways non-renewable resources can be conserved?

9. If further practice seems like a good idea, or if students are excited about the research, ask students to trace one or two other items back to their origins as homework. These items may be additional ones from their lists, those that they couldn't trace during class or simply something they are curious about -- perhaps their favorite objects.
    Have the students display their findings on a poster or bulletin board. Non-renewable resources can be written in one color, renewable ones in another. For example, working in groups of three or four, students can consolidate the information on the poster by diagramming products from their common resources in a branching tree formation such as the one below. (This would take an additional class session.)

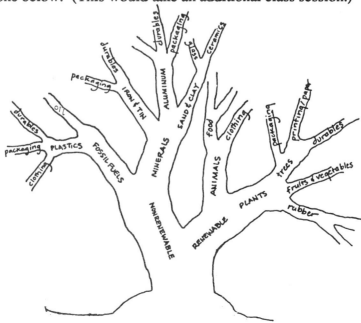

Activity 4

THINKING ABOUT *THE LORAX*

Objective

In this activity, Dr Seuss's *The Lorax* will lead students into exploring the environmental effects of manufacturing and understanding the role of consumers in production. They will examine how controversy grows out of environmental concerns and recognize the complexity of decision-making about natural resources.

Note

The issues in this activity can be addressed at various levels of complexity. See "Supplementary Information" following procedure 2 for suggestions.

Suggested Previous Activity

3. Identifying Natural Resources

Time

45 minutes for story and discussion
1 homework assignment
30 minutes for discussion of article, or 2 45-minute periods for role play
30 minutes for wood substitution activity (optional)

Vocabulary

clear-cutting	natural resource
monoculture	selective cutting

Materials

copies of *The Lorax* by Dr. Seuss, or one copy to be read aloud, or *The Lorax*
   video available from either the EPA or video stores
copies of Handout 3, "A Boy Sides with Dr. Seuss's Lorax..." or a summary of it
copies of "Roles" on Handout 4, cut and separated
video camera (optional)

Activity Procedure

1. Read *The Lorax* aloud in class, view the video, or assign the book to be read for homework. Certain students may feel they are too old for Dr. Seuss. Therefore, before reading, ask the class if any of them have read *The Lorax*. What do they remember?

Suggest that they think of the ways a younger child would react to the book as they themselves read it in class.

2. Discuss the story using the questions below or others that you create.
 * Why did the Once-ler cut down the Truffula trees?
 * Why do the Bar-ba-loots, Swomee Swans, and Humming-fish have to leave?
 * What kinds of problems does the Thneed factory cause for the environment?
 * What could the Once-ler have done to minimize his factory's effect on the environment?
 * A Thneed is defined as a "Fine-Something-That-All-People-Need." What are some "thneeds" that we think we need?
 * Do you think that Dr. Seuss has represented all of the members of logging industry?
 * What are the various ways each character has responsibility for the destruction of the environment in the story? (How are the people who bought "thneeds" responsible?) Is any character free of responsibility?
 * What do you think the Lorax's message "UNLESS" means?

Supplementary Information for Teachers

There are a number of issues that could make the scenario depicted in *The Lorax* more complicated. We offer two of them here as a supplement to be refered to as desired, with related suggested questions for students. These issues are the use of natural resources for medicinal purposes, and the intentional growing and harvesting of trees.

A. The use of natural resources for medicinal purposes: a case study of Taxol

What if a thneed was not just an extra something that every one wants? What if a thneed could cure cancer? Recently it was discovered that the bark of a 100-year-old Yew tree (Latin name *Taxol)* found in the Pacific Northwest is effective in treating certain kinds of cancer. The problem for many environmentalists is that it takes six 100-year-old Yew trees to treat one cancer patient. It is likely that at least 200,000 patients a year would seek the drug. Environmentalists are concerned that the spotted owl, who lives in the Yew, will become extinct along with many other species, not to mention the potential loss of the Yew tree itself.

So far, Taxol (the name given to the new drug after the Latin name of the tree) has only been used on cancer patients for whom other treatments have failed About a quarter to a third of the people who have tried the drug have responded favorably. Tumors have gotten smaller, and some have disappeared for months. No one has been cured by Taxol yet, but it has enabled patients to live longer and with less discomfort (Kolata, 1991)

 * Do you feel there is any difference between cutting down trees to make toys, furniture or other items, and cutting them to make medicines for life threatening diseases? Why do you feel this way?
 * What would you and/or the Lorax think about killing trees to provide a medicinal cure for a sickness that kills 10 people per year? What about 10,000 people per year?
 * What if the trees were discovered to have bark that could be made into a pain-reliever ? acne cover-up? Would you feel differently if the trees were an endangered species?

Fortunately, we do not have to answer the questions specifically about Taxol. A drug company has found ways to extract the necessary ingredients to make Taxol from the needles of the Yew tree, so the trees will not have to be killed to manufacture the anti-cancer drug. But that may not be true of other natural resources found to have medicinal value.

B. Can logging be done intentionally and responsibly?

How accurate is Dr. Seuss's portrayal of the logging industry? In *The Lorax*, it seems that loggers find forests and cut down all the trees. Actually, there are several practices that vary among logging companies. Some companies own land and cut down trees that they have grown specifically for harvesting. Then they replant, usually with one kind of tree. This creates a "monoculture," which limits the diversity of animals that could live in that environment.

Sometimes, even if companies own land for growing and harvesting, consumer demand is so great that companies must find more trees to cut beyond those on their own land. For this purpose they frequently turn to the national forests. The processes used to harvest these trees are "clear-cutting" of large areas, clear-cutting of many small areas, or "selective cutting." Clear-cutting of large areas is an efficient means to obtain a large number of trees. However, it can destroy animal populations dependent on that particular type of forest. Clear-cutting many small areas may be less efficient but is also less destructive to local animal populations. Selective cutting, once thought of as the most environmentally sound method, is now thought to require the most energy consumption for the number of trees obtained: it also causes extra damage to the forests due to the moving of equipment and people in the process of obtaining the selected trees.

*Should companies be able to do what they want on land that they own? Is intentionally growing trees for harvesting the best way to provide trees for wood products?
*What method did the Once-ler use to cut the trees? What other methods could the Once-ler have used to harvest the Truffula trees?

3. Read or summarize the article from *People* magazine in class, or assign the article for homework. After reading the article, discuss the controversy using the following questions, or do the role play described below.

Discussion Questions

* Where do most families in Laytonville work? Why are they so strongly affected by *The Lorax*?
* Why did one mother say her child had to choose between "Dr. Seuss and Daddy"?
* What do you think the town should do to bring the various factions together?
* Does Dr. Seuss seem to be saying that loggers are bad, or do you agree with his statement that he is not saying that?
* Do you think the book should be banned? Why or why not?

Role play

After reading the story and the article, students can create a meeting of a town council that allows for all sides to be heard and then try to come to an agreement.

Handout 4 outlines some roles that could be used, as well as questions to help students define each person's attitude. If you prefer, assign roles without using Handout 4. Encourage students to figure out their characters' concerns and argument tactics for themselves.

This activity can be done by assigning groups of three students to each role. The roles and corresponding questions should be cut up and separated so that each group sees only those questions that pertain to them. Each group formulates a position based on the questions provided and other ideas students may have. The groups can then appoint one group member to play the role at the meeting, or three meetings can be held so that all students have the opportunity to play their roles. The teacher may want to add additional characters depending on class size. New characters could be more neutral on the issue.

If a video camera is available, students may want to dress up and have theeeir performance videotaped.

The following questions can be used for a discussion after the role play:

* Was a decision about *The Lorax* made? If not, why not? If yes, what was the decision, and how was it made?
* Was each speaker truly heard by other "citizens"?
* What was difficult or easy about playing your role?
* What did you like or not like about playing your role?
* Is there anything you wish you had done or said differently in playing your role?

4. Optional follow-up activity: Have students, working in small groups, make lists of all the things they use in the classroom that come from trees. In the same small groups, or as a whole class, ask students to decide whether all the items made from trees are necessary. Could anything be eliminated or substituted? How would the substitution affect the environment? the school budget? the classroom setting? (Example: Consider replacing wooden desks with plastic ones.)

A BOY SIDES WITH DR. SEUSS'S LORAX, AND PUTS A TOWN AT LOGGERHEADS *

The trouble began the day Sammy Bailey came home from school last spring . The Laytonville, Calif., second grader had just finished reading *The Lorax* by Dr. Seuss, the sad tale of a fuzzy little creature who loses his forest home when the greedy Once-lers cut down all the Truffula trees. A troubled and thoughtful Sammy had taken the story's lesson to heart. "If you cut down a tree," he told his father, Bill Bailey, "then it's just like someone coming in and taking away your home."

Another parent might have been touched by his child's sensitivity; Bailey was not. The owner of a logging supplies mail-order business, he was incensed by what he saw as a flagrant attack on the livelihood of Laytonville, a tiny (pop.1,096), single-industry lumber town 150 miles north of San Francisco. Rounding up support from other outraged parents, Bailey, 46, and .his wife, Judith, 42, asked the local school board in September to remove *The Lorax* from the second-grade required-reading list. "Teachers....mock the timber industry, and some of our kids are being brainwashed," screamed Bailey's full-page ad in the local weekly. "We've got to stop this crap right now!"

The skirmish quickly mushroomed into a cause celebre when some teachers and other townsfolk, viewing the Baileys' campaign as a threat to academic freedom, responded with equal passion. The issue came to a head early this month as 150 parents, teachers and kids gathered for a boisterous showdown with the schoolboard in the lunchroom of the Laytonville Elementary School. After some 30 pro-*Lorax* speakers and one pro-logger argued their case, the board voted to keep the 1971 children's classic on the reading list, at least for now. But board members also scheduled a meeting next month to determine whether that whole list should be preserved--or shelved.

The only certainty is that the dust has not yet settled in Laytonville, where a feud has long been simmering between loggers and conservationists Last year a group of Earth First! environmentalists set up a blockade for three days outside town, preventing the timbermen from cutting down old-growth trees and destroying the habitat of the rare spotted owl. "The spotted owl is dependent on those trees for survival," says Sierra Club representative Bill Arthur. "As goes the owl, so goes the health of the forest."

Now, with *The Lorax* and its conservationist message bringing the battle into the classroom, Laytonville's loggers are feeling besieged. "Our industry is under attack," says Art Harwood, 36, manager of a sawmill that employs 300 people in the neighboring Branscomb. Now, he says, "(the environmentalists) are trying to hang the ozone and the rain forest on us, and it's easy to take that *Lorax* and use it against us too." The father of

* Reprinted by permission of *People Magazine*

two children, Harwood says the Seuss book isn't the only problem at the town's schools; he cites an assembly at which students were taught environmental theme songs with lyrics such as "These are trees that live so long, it's really a shame to cut them down." He is also disturbed that sawmill representatives weren't invited to Career day last April at Laytonville High. Judith Bailey's complaint is more personal. "I don't mind the *Lorax* being taught to a child at an age when they don't have such a black-and-white view of good and bad," she says. "But it's stressful on the child when he has to choose between Dr. Seuss and Daddy."

As for Theodore Geisel, 85, better known as Dr. Seuss, he says he wrote the book not about logging but about the cause of conservation in general. The La Jolla, Calif., author of 47 rhyming works, including *The Grinch Who Stole Christmas* and *The Cat in the Hat*, says *The Lorax* "is about people who raise hell in the environment and leave nothing behind. I'm not saying logging people are bad. I live in a wooden house and sit in a wooden chair."

Meanwhile, to 9-year-old Sammy Bailey, the issues of academic freedom, logging and the fate of the spotted owl no longer seem paramount.

Though he is aware that his parents have been labeled book banners, he is more concerned now about his third grade studies, the care of his pet rabbit, Silver, and his chores as water boy for the local high school football team. Sammy won't say any more about *The Lorax*, but class-mate Jenni Cothern, 8, thinks she understands what Dr. Seuss had in mind. "The lesson was," she says, " ' Don't chop all the trees at one time 'cause the animals will have no place to live.' " Or as the Lorax counsels, "Plant a new Truffula. Treat it with care. Give it clean water. And feed it fresh air." ——— **Ron Arias, Liz McNeil in Laytonville**

Handout 4

ROLES

Town Councilors (5):

The role of the council is to make sure each speaker is heard and to try to reach an agreement about banning or not banning the book *The Lorax* in schools. During the meeting, the members of the council are in charge of calling on people to speak. While the other participants are preparing, the councilors should come up with questions that they want to ask each of the participants listed below. After all views have been expressed, the councilors leave the room to confer and then return to present their final decision.

The other participants include the owner of a logging company, a local citizen (not a logger) who supports academic freedom, a teacher, a Sierra Club member, Dr. Seuss, the wife/husband/partner of a logger, a furniture manufacturer, and a local citizen (not a logger) who supports the book ban.

Owner of Logging Company:

What is your major source of income?
From your point of view, what role in the logging industry do consumers of wood and paper products have?
Why do you think *The Lorax* is harmful to young children?
What contribution does your company make to the community?
What would happen to the community if your company closed?

Local Citizen (not a logger), Supporter of Academic Freedom:

How do you feel about censoring books?
Why should *The Lorax* be read in Laytonville schools?
What should students be taught in school?
How can academic freedom be retained while addressing the concerns of the loggers?

Teacher:

What is academic freedom and why is it important?
How can you be sensitive to the issues of loggers and still read *The Lorax*?
How would your job be different if every book that offended someone were censored?
Should children be exposed to real life controversy in school?
Why is it important to read *The Lorax*?

Sierra Club Member:

How are trees important to wildlife?
What message from *The Lorax* is important to you?
How do you feel about loggers?
What are your concerns about how people effect the environment?
What are your concerns about logging?

Dr. Seuss:

Why did you write this book?
How do you feel about logging?
What responsibility do consumers of wood and paper products have in this controversy?
What are your environmental concerns?
What did you want young people to think about when they read this book?

Wife, Husband or Partner of a Logger:

What are your concerns about your child's education?
What is the role of the school?
What conflict is created for your family by reading *The Lorax*?
How could the concerns of loggers be presented in school?
How do you feel about literature that influences your child with a political message?

Furniture Manufacturer:

In what ways is it hypocritical for people to buy furniture from your store while opposing logging?
Why do you think *The Lorax* is unrealistic?
If Dr. Seuss is so concerned about the environment, why wasn't the book printed on recycled paper?
What is the responsibility of environmentalists who oppose logging but buy wood products?
Why are the images in *The Lorax* bad?
What should furniture be made from, if not wood?

Local Citizen (not a logger, Supporter of Book Ban:

What will happen to the community if the logging business closes?
Why don't people recognize that we use and need wood products?
Why should political perspectives be introduced to young children (or not be introduced to them)?
Why is reading persuasive books harmful to the community?
Why are the images in *The Lorax* bad?

Local Resident, Member of Earth First!

How does putting up blockades so loggers can't do their work help the community in the long term?
Why is cutting down all the trees bad for the logging industry <u>and</u> the environment?
Why do humans sometimes have to make sacrifices to save the environment?

Activity 5

CONSTRUCTING A MINI-LANDFILL

Objective

Students will each make a miniature landfill and be able to explain how a landfill works. They will discover what happens to biodegradable and non-biodegradable materials in a landfill. Students will also understand what "leachate" is and how it is created.

Note

Although it is not possible to fully duplicate landfill conditions in a carton, making mini-landfills will give students an idea of how landfills work and what happens to trash in them. Students will be working in pairs on two mini-landfills. They will be experimenting with a variable in their own pair of landfills. To demonstrate what conditions help break down trash or prevent liquid from leaching out of the landfill, you may have one or more pairs of students try each of the variations described below.

Although decomposition would occur more rapidly in a landfill exposed to sunlight and air, most landfills are compacted and buried for the sake of space, health, and aesthetics. In this experiment, therefore, the mini-landfills are not exposed to sunlight and air. Another approach would be to expose some of the mini-landfills to light and air and to compact and cover others, noting the differences. It is important, however, for students to understand that in actual landfills, trash is rarely left exposed.

Holes are placed in the mini-landfill cartons so that when landfills are watered (rain), some water leaches out. This liquid represents leachate in a landfill. Many landfills today are lined with clay or plastic, which is supposed to contain the leachate. However, older landfills are not lined, and newer ones do not always demonstrate foolproof containment of leachate. One pair of students can add a clay liner to the bottom of the landfill and compare the resulting leachate to those landfills without the clay.

Laboratory research shows that material decomposes more quickly when greater quantities of water are added. You might want to have two pairs of students make landfills from the same ingredients and then vary the amount of water each pair uses. In this way the students can determine if more water promotes more rapid decomposition. (Of course, more water will also create more leachate.)

Time

    45 minutes to make the mini-landfills
    1 month to tend them and allow for decomposition
    45 minutes to find results and discuss

Vocabulary

biodegradable   non-renewable resource
landfill   organic
leachate   recyclable
non-biodegradable   renewable resource

Materials - needed for **making** each landfill (for each student)

1 half-gallon cardboard milk or juice container
1 small plastic container (8 oz., the kind margarine comes in, strong enough to hold up the full half-gallon carton)
1 large plastic container (32 oz., the kind margarine or ice cream comes in)
1 heavy object (such as a brick or rock) used as a weight to fit inside the top of the carton
soil - enough to fill the carton 3/4 full (**not potting soil** as potting soil has been sterilized and therefore lacks the microorganisms necessary for decomposition)
clay - enough to cover the bottom of one mini-landfill with a half-inch layer
garbage (a mixture of food and non-food), cut into small pieces (try to include a commercial item said to be biodegradable)
scissors
1 measuring cup
water
2 copies of Handout 5, "Mini-Landfill Chart" for each pair of students

Materials - needed for **analyzing** landfills at the end of one month

mini-landfills, now one month old
newspaper
popsicle sticks or spoons
2 copies of Handout 6, "Mini-Landfill Analysis Worksheet" for each student

Activity Procedure

1. Have students pair off to create two mini-landfills according to the instructions in step 3 below.

2. As a class, have students decide what variable they would like to experiment with. Each pair should then make one landfill the normal way and one landfill altering the variable chosen. Suggested options include
   * exposing some landfills to light and leaving others in the dark
   * adding a clay liner to some of the landfills and not to the others
   * watering the landfills in different amounts

Note: The landfills are prone to falling over, so make sure the students' constructions are stable.

3. Explain to students the following steps to make their landfills:

   A. Open the top of the cardboard milk or juice carton so you will have room to add materials.

   B. As the landfill is made, fill out the Mini-Landfill Chart, listing the garbage in each layer. (They will fill out the last row one month later.) Use the second Mini-Landfill chart for the mini-landfill with the variation.

   (For clay-lined landfills, see step E before going to step C.)

   C. Add a layer of soil, then a layer of garbage, then soil, garbage, soil, garbage, ending with a layer of soil. The final product should consist of four layers of soil alternating with three layers of garbage. Pack down each layer tightly after soil is added. Soil layers should be about two times as thick as garbage layers. (For a clay-lined landfill, flatten the clay on the bottom and sides of the carton before adding the soil and garbage. The clay should form a half-inch layer across the bottom and about an inch up the sides, so that the holes are covered.)

   D. Place the brick on top of the last layer.

   E. Using the point of the scissors, make small holes on the sides near the bottom of the carton as shown. (For clay-lined landfills, make the holes before putting the clay or the garbage into the carton. Otherwise the scissors may puncture the clay.)

   F. Place the small container upside down inside the large container. Then place the cardboard carton on top of the small container as shown. In this way moisture will be able to drain from the carton into the large container. (Make sure the construction is stable. The middle container should be large and strong enough to support the milk carton. Filling it with dirt can make it stonger.)

4. During the month have students water the landfills every other day, using the measuring cup. Use approximately 1/4 cup of water each time. If the landfill becomes saturated, water less often. If the top layer is drying out, water more often.

5. The following activities should be completed before analysis of the landfills:

      2. Changing Times, Changing Waste
      3. Identifying Natural Resources
      7. Natural Cycles

6. At the end of the month have students remove the soil layers and record what happened to the layers of garbage on the Mini-Landfill Chart and Mini-Landfill Analysis Worksheet. Students should place newspaper on the table and remove contents of the landfill onto the newspaper layer by layer (as much as possible), using the scoop or spoon. Students can use popsicle sticks to sort through contents.

7. Discuss in small groups or with the entire class the worksheet questions, the students' discoveries, and the implications they think their findings have for actual landfills. (Regarding question 6 on student worksheet: Spaces occurring in a landfill result from decomposing material within the landfill. After a landfill is covered, these small spaces can cause some shifting in the land, posing a problem for building on the land.)

    * How might their mini-landfills differ from a real one? How are they similar?
    * Given their findings, how do landfills help trash disappear?
    * What are the shortcomings of a landfill?

Names _____

_____

MINI-LANDFILL CHART
Variation:_____

	Bottom layer	Middle layer	Top layer
List items of garbage			
Which items are organic?			
Which items were made from a renewable source?			
Which items are recyclable?			
Which items do you think are biodegradable?			
Write a prediction of what this layer will look like at the end of the month.			
Write a description of what this layer actually looked like after one month.			

MINI-LANDFILL ANALYSIS WORKSHEET

1. Which items decomposed the most?

2. Which items decomposed partly?

3. Which items did not change at all?

4. What characteristics are shared by the items that decomposed quickly?

5. In what other ways might we dispose of the items that did not decompose?

6. Sometimes spaces form around decomposing waste items. What do you think causes this? If this happened in an actual landfill, how could the spaces affect the use of the land after the landfill is covered and closed?

7. During the month, how much liquid, or "leachate," came out of the landfill into the container?

8. If one pair of students used a clay liner, was there a difference in the amount of leachate that came out of that landfill compared to that from the non-clay lined landfills?

9. What do you think happens to leachate in an actual landfill?

10 Do you think a clay liner would help contain leachate? What problems might occur?

Activity 6

COMPOSTING

Objective

Students will understand the differences between compost piles and landfills and consider the positive and negative aspects of composting. Through experimentation students will learn how to make an efficient compost pile and will learn about the recycling of organic waste.

Note

This activity can be done in two ways. (1) Make one or two large compost piles outside, or (2) use terrariums to make compost piles inside. The number of piles made depends on your resources (the number of terrariums for the indoor piles and the number of containers and space for the outdoor piles). The more piles that are made, the more experimentation is possible.

Suggested Previous Activities

1. How Much Trash Do You Throw Away in a Week?
2. Changing Times, Changing Waste
5. Constructing a Mini-Landfill

Time

45 minutes to set up the compost pile
45 minutes after 6-8 months for follow-up discussion

Vocabulary

actinomycetes	fungi
aerate	humus
aerobic	nutrient
bacteria	organic
compost	primary consumers
decompose	secondary consumers
decomposers	tertiary consumers
food web	

Materials

Handout 7, "Necessary Ingredients for a Compost Pile"
Handout 8, "What is in your Compost Pile"
Handout 9, "Temperature of Your Compost Pile"
graph paper
containers (see below)

fine nylon mesh (see below)
soil (**not potting soil** which is sterile)
organic wastes (see below)
200 redworms (see below)
thermometer
pitch fork (for outside compost) or trowel (for indoor compost) for turning
        compost pile

Container: For use indoors, a plastic bin, 1 foot by 2 feet by 3.5 feet, is recommended. The plastic bin should have 8-10 holes drilled into the bottom for drainage. The bottom should be lined with a fine nylon mesh to prevent the worms from escaping. For use outside, a compost container can be designed in many different ways and with many different materials. (See pictures below.) They can be made in cages, blocks, pits, drums, or garbage cans. In most cases the design and materials used depend on the resources and needs of the user. Commonly used materials include chicken wire and wooden slats. It is also possible to buy commercial composters. A container is not necessary for outdoor composting, but some people think it looks nicer and it can prevent animals from getting into it. The outdoor container should be at least 3 feet by 3 feet by 3 feet to hold in the heat, and it should not be higher than 5 feet. If a compost pile is much taller than 5 feet, it can get too compacted, and not enough air will be available to the decomposers. Because a compost pile needs air, be sure the container used has openings to allow air to get in and that it is structured in such a way that it can be turned. (Pit compost bins can be hard to turn with out straining your back.) There is no need to have a floor on the container.

Organic wastes: These wastes include food (not meat products or greasy or oily foods, as they can attract rodents and smell bad), leaves, grass clippings, manure, hay, etc. It is not recommended to use either sawdust or ash, because many wood products have been treated with toxic chemicals. (Most lumber has been treated with toxic chemicals, which could become part of your composted soil.) Be sure to have a variety of organic materials.

Red worms: Red worms are recommended because they consume the most waste for their size in the quickest amount of time compared with other worms.

Supplementary Information for Teachers

Composting has been around since the origin of life. When the first primitive plant or animal died and was decomposed, this was composting. Human involvement with composting probably dates to when people first cultivated food and animals. The oldest existing reference to the use of manure used in agriculture was found on a set of clay tablets of the Akkadian Empire which existed in the Mesopotamian Valley one thousand years before Moses was born. Composting was known to the Romans and the Greeks. The Bible has many references to "dung" and "dunghills" which were used on the fields (Martin and Gershung, 1992).

Today composting is being encouraged by state environmental protection agencies because of the increasing concern over the vast amount of trash this country produces. Composting is actually a form of recycling. Recycling often refers to glass bottles, aluminum cans, and newspapers. However, the 26% of trash that is yard and food waste and is generally thrown away could be recycled. Organic waste such as food, leaves, grass clippings, and wood scraps are biodegradable and can be recycled by composting.

Composting is a controlled process of decomposition of organic material. Decomposition is the biochemical process by which bacteria, fungi, and other microscopic organisms break down organic "wastes" into nutrients that can be used by plants and animals (Hallowell, et.al., 1989). Decomposition occurs naturally whenever a leaf falls to the ground or an animal dies. Decomposition is nature's way of recycling many nutrients.

Composting creates optimal conditions for decomposition to occur. These conditions include the right balance of air, water, temperature, soil, the presence of decomposers, and food (for the decomposers). Air is commonly added to the compost pile by turning it on a regular basis. Water is obtained naturally or is added by hand to keep the pile moist. Soil contains the decomposers. Many decomposers -- the microorganisms, worms, centipedes, etc. that work to break down organic waste -- need water to live and reproduce. The temperature of the compost pile is controlled by the volume of composted material and the kinds and amounts of material put in.

The food for the compost pile comes from the materials added. Two elements essential for microorganisms to live are carbon and nitrogen. All organic material contains carbon and nitrogen in different amounts. The microorganisms need carbon for energy and nitrogen to reproduce. When the carbon to nitrogen ratio is approximately 30 to 1, the microorganisms are most productive. Various carbon to nitrogen ratios are listed on Handout 7, "Necessary Ingredients for a Compost pile." Ask students if they can guess or generalize as to which kinds of materials are high in carbon and which are high in nitrogen. A general rule of thumb, according to a Department of Environmental Protection (DEP) of Massachusetts composting expert, is that brown organic substances are high in carbon and green organic substances are high in nitrogen. Dry, woodsy materials like leaves, straw, and cornstalks are high in carbon. Green weeds , grass clippings, and animal manure are high in nitrogen. Green leaves that are left out to dry turn brown and become high in carbon (DEP, 1993).

According to the DEP of Massachusetts the following items should not be added to a compost pile: cat and dog manure (can carry harmful diseases), "diseased plants, weeds that have gone to seed, and seeds that spread by rhizomes, such as morning glory, ivy and quack grass as they may not be killed during composting and could take root wherever the compost is used...." (DEP, 1993).

How the material is put into the compost pile can affect how quickly it decomposes. Alternating layers of high carbon materials with high nitrogen materials, with small layers of soil in between, is recommended. Each layer should not be more than 6 inches thick. Since the decomposers that do the work in a compost pile live and work on the surface of organic material, the more surface area the microorganisms have to work on, the faster the materials will decompose. To draw an analogy, a block of ice in the sun will melt faster when it is broken down into smaller pieces. Challenge students to think about how they can increase the surface area of the material they add to the compost pile. (For example, shredding materials to 1-inch pieces before adding them will increase the surface area.)

The work of breaking down organic wastes into soil is done by organisms that live in the soil. Bacteria, fungi, and actinomycetes start the process of decaying organic matter. They are called the primary level consumers. (Additional primary consumers are given in the chart, "Food Web of the Compost Pile.") The secondary consumers, such as the protozoa, beetles, mites, springtails, and flatworms, eat the primary consumers. Finally, the tertiary consumers, the centipedes, eat the secondary consumers. The food web of the compost pile transforms the organic wastes into a nutrient-rich humus that is excellent for improving soil quality and plant growth.

Composting is probably the most beneficial way to dispose of your yard waste. It is inexpensive, can improve the quality of your soil, and can help reduce the amount of trash sent to landfills and waste-to-energy facilities.

Activity Procedure

1. Ask students what they do with their yard waste and/or leftover food at home. Some may throw food "away"; others may compost. If any students compost ask them to explain what composting is. If no one does composting, ask the students what else can be done besides throwing such waste away. Ask what happens to the leaves in a forest or to a tree when it dies. (Leaves and trees decompose over time, forming soil.) Ask if any students have heard of composting.

2. Decide whether you are going to make one compost pile as a class or whether you have the resources and inclination to have students work in groups of four to make their own compost piles. Ask the class or groups to decide how they are going to make their compost pile based on the information in Handout 7, "Necessary Ingredients for a Compost Pile."

3. Have students answer questions on Handout 8, "What is in Your Compost Pile?"

4. Each group or the class should keep track of the temperature on Handout 9, "Temperature of Your Compost Pile." They can make line graphs using their data.

5. If the compost pile is indoors, place it where it will be at room temperature and not in direct sun.

6. Hold a class discussion just after making the compost pile(s). Discuss with the class the differences between a landfill and a compost pile. (Landfills are always dug into the ground, have all kinds of trash, and are not aerated. Compost piles are usually built on

top of the ground, though they also work when dug into the ground; they use only organic wastes and create a product that can be used.)

* What are the advantages and disadvantages of composting? (Advantages: decreases waste stream by 26%, provides nutrient-rich soil, reduces number of plastic garbage bags needed. Disadvantages: time consuming to construct the bin and to take care of pile)

* Where did the idea of composting come from? Is it a new idea? (No; see background information.)

7. Return to this discussion after six to eight months. If the class made different compost piles, discuss which compost piles were most successful and why.

* What are the optimal conditions for a compost pile?
* Where does composting occur naturally?
* Does composting reduce waste? If so how? If not, why not?
* How is composting related to recycling?
* Is a landfill a giant compost pile? Why or why not?
* Should towns require composting? Why or why not?

Alternative Activity

This activity can be presented as a challenge to students to see who can get a certain amount and kind of organic material (provided by the teacher) to decompose the fastest. The faster the compost material decomposes, the sooner one is able to use it to nurture plants. Follow the directions for the activity as written, except make sure that all students start with the same amount and kind of organic wastes.

Suggested Follow-up Activity

Is composted soil any better than regular soil? Have students plant flowers or vegetables in two different soils, the composted soil and soil from outside. Have students evaluate the two different soils by comparing growth rates and the taste of fruits and vegetables from the plants.

NECESSARY INGREDIENTS
FOR A COMPOST PILE

Soil contains the decomposers. Many of these decomposers are microscopic, such as bacteria, fungi, actinomycetes, and protozoa. These organisms change the chemistry of the organic wastes and are referred to as the chemical decomposers. The physical decomposers are small but can be seen without a microscope. They include mites, millipedes, centipedes, sowbugs, snails, slugs, beetles, ants, flies, and worms (Martin and Gershung, 1992).

Nitrogen is needed by the organisms responsible for decomposition for rapid and thorough decomposition. Nitrogen is found naturally in many organic wastes, such as manure and green grass clippings, fresh green plants, and food wastes, as well as in many commercial fertilizers.

Carbon is needed by the decomposers to be used in conjunction with the nitrogen. The decomposers convert the carbon to energy. Materials that are high in carbon include dry woody materials, cornstalks, dried leaves.

Organic waste includes items such as leaves, food scraps, and grass clippings. Wastes should be varied, including materials with both carbon and nitrogen. By alternating layers of high-carbon and high-nitrogen materials, you can create good environmental conditions for decomposition to occur. Nitrogen and carbon are used by microorganisms for energy and growth. The microbes are most active when the ratio of carbon to nitrogen is about 30:1. Too much carbon will mean that the decomposition process with take longer. Too much nitrogen will cause the compost pile to smell bad. Below are the carbon to nitrogen ratios of some organic materials.

food waste	15:1
wood	700:1
sawdust	500:1
straw	80:1
fresh grass clippings	19:1
dead leaves	60:1
fruit wastes	35:1
rotted manure	20:1
cornstalks	60:1
alfalfa hay	12:1
newspaper	170:1

Remember, for the fastest composting, you want to have a carbon to nitrogen ratio of 30:1

Handout 7 continued

(Newspaper ink is carbon and does not pose a health problem as long it is only black ink. Colored ink is hazardous, as are other types of paper, which are often made with bleaches and other chemicals. The leaves referred to are not freshly fallen leaves but rather leaves that have been on the ground for several weeks. Freshly fallen leaves would have a lower carbon to nitrogen ratio.)

Worms eat the waste, helping to break it down; make droppings, which enrich the soil, tunnel through and aerate the waste; and eventually die and become part of the compost.

Water is necessary for the normal functioning of life. Too much water in a compost pile may make it soggy and slow decomposition by reducing needed oxygen. Compost organisms need a steamy environment. The pile should be moist, like a wrung-out sponge, but not wet. To test for the correct amount of water, squeeze a handful of the composting material; it should feel damp, and no more than one drop of water should come out.

Air is necessary for the decomposers to thrive. The aerobic creatures (fungi, bacteria, small insects, earthworms, and other organisms) who do the work of composting need a good supply of oxygen, since most biological processes require adequate amounts of this gas. Turning the pile frequently is the best way to maintain good aeration throughout the pile.

Decomposition takes **time**. To speed up decomposition, aerate your pile every few days; otherwise, just leave it and wait.

Heat is produced by chemical reactions resulting from increased biological activity that occurs during decomposition. Heat helps sanitize compost by killing certain unwanted organisms, such as weed seeds, pathogens, and harmful insect larvae.

Mass is important. To generate enough heat for optimal decomposition, the pile must contain at least one cubic meter of organic material. If there is not enough heat, the compost pile will break down, but it will take much more time. Thus, the temperatures generated in an aquarium compost pile may be different from those generated in one that is larger depending upon the amount of organic material used.

FOOD WEB OF THE COMPOST PILE

ENERGY FLOWS IN THE DIRECTION OF THE ARROW. 1° = FIRST LEVEL CONSUMERS
2° = SECOND LEVEL CONSUMERS
3° = THIRD LEVEL CONSUMERS

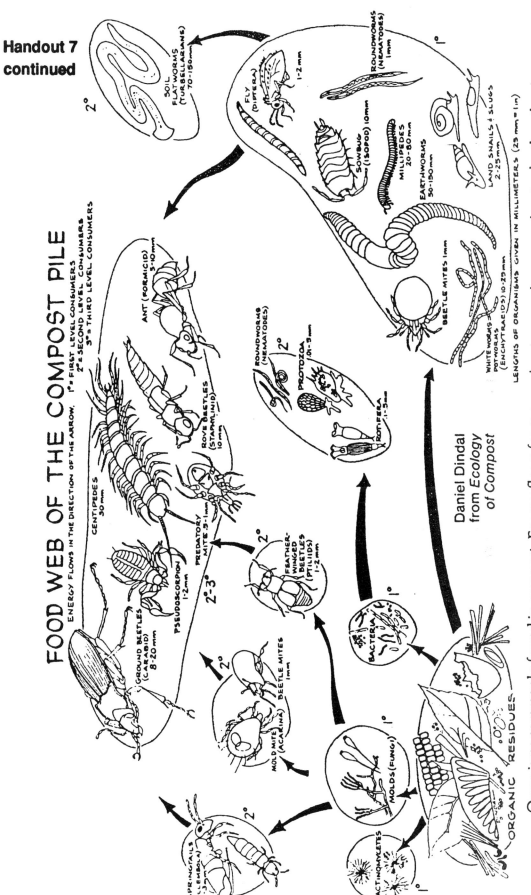

Daniel Dindal
from *Ecology of Compost*

Organisms commonly found in compost. Energy flows from organism to organism as one is eaten by the other in a natural recycling system.

Name _____

WHAT IS IN YOUR COMPOST PILE?

1. Describe how you made your compost pile. What did you put in, and how much of each item did you put in? Why did you include the ingredients you did?

2. How are you going to care for your compost pile? How often will you give it water? How often and by what method will you give it air?

3. Draw a picture below of what your compost pile looks like now. Then draw what you think it will look like in six months.

     NOW   **IN SIX MONTHS**

4. Draw a picture of what your compost pile actually loolked like after six months.

Name _____

TEMPERATURE OF YOUR COMPOST PILE

Directions: Take the temperature of your compost pile every week. Try to take it at the same time and at the same depth each week. Record your temperatures below.

Graph the temperatures on graph paper when your experiment is done.

DAY	TIME	TEMPERATURE

Activity 7

NATURAL CYCLES

Objectives

This activity focuses on the natural cycles that produce many life-sustaining resources and examines how humans effect these cycles. In Part A, students will learn where water supplies come from and how the water cycle works. In Part B, students will learn what leachate is and how it effects the water supply. Through Part C, students will understand both the oxygen and nutrient cycles.

Suggested Previous Activities

1. How Much Trash Do You Throw Away in a Week?
3. Identifying Natural Resources
5. Constructing a Mini-Landfill
6. Composting

Part A: Demonstration of the Water Cycle

Time

20 minutes

Vocabulary

aquifer	precipitation
evaporation	transpiration
groundwater	watershed

Materials

1 hot pot, or electric burner and small pot
1 teaspoon salt
1-2 cups water
2 metal trays
15-20 ice cubes
1 copy of Handout 10 "The Water Cycle" for each student

Supplementary Information for Teachers

The following information is reprinted with permission from "*Water Watchers*," produced by the Massachusetts Water Resources Authority and The Museum of Science, Boston.

"Land surfaces are divided into natural subdivisions called watersheds. The rainwater or snow that falls in a watershed will drain into streams, rivers, and eventually the ocean. All land surfaces are in one watershed or another. Watersheds are defined by the highest ridges at the top of hills or mountains.

The rain or snow that falls in a watershed can do several different things: 1) it can travel over the surface of the land as runoff forming streams as it goes; 2) it can soak into the ground to become groundwater; 3) it can be taken up by the roots of plants and returned to the air by a process called transpiration; or 4) it can evaporate into the air from open water surfaces.

Runoff is the water that forms streams and rivers. Groundwater moves around under the ground in a manner similar to rivers on the surface, but much more slowly. The areas where ground water flows are called aquifers.

Evaporation returns water to the air as vapor. It takes place from open surface areas of water like the ocean. The larger these areas are, the more water vapor goes into the air.

Transpiration is a process whereby plants release water vapor from their leaves. This water was previously taken up by the roots from water in the soil. (Transpiration is a process similar to perspiration in humans, in which water evaporates from the surface of the skin.)

People get their water supply by tapping either groundwater or the runoff. Groundwater is tapped by drilling a well to reach down into an aquifer, and then pumping the ground water to the surface. Runoff is trapped by building a dam to stop the flow of a stream or river. The dam creates a reservoir, a human-made water storage lake. In some places people draw runoff water directly out of a stream or natural lake."

Activity Procedure

1. Place the hot pot, or electric burner and small pot, on top of one of the trays. In the pot, put one or two cups of water and one teaspoon of salt to represent ocean water. Allow water to come to a boil.

2. Put the ice cubes on the other tray. Place this tray over the steam coming from the boiling salt water. As the water evaporates, becoming steam, it hits the cold tray and condenses, forming water droplets. If enough droplets form, it will "rain" into the lower tray.

3. Have one student taste the evaporated water by collecting one drop on his or her finger. Does it taste salty? (No. Neither does water that evaporates from the oceans.)

4. Go over the water cycle handout with students and explain it in relation to the demonstration.

5. Hold a class discussion using the following questions:

* Why are ice cubes used on the tray? What do the ice cubes represent? (The ice cubes cool the tray, thereby cooling the steam and causing it to condense and form water drops. The cubes represent the cooling temperatures as evaporated water moves upward in the earth's atmosphere.)

* Compare this demonstration to the actual water cycle. How is it the same? How is it different? (Water in oceans and lakes evaporates, rises, condenses, and returns to the land by some sort of precipitation. In this demonstration, the steam is like evaporating water. The tray with ice cubes represents the cooling atmosphere. Droplets form and fall back to the "land." This demonstration is different from the actual water cycle in several ways. For example, precipitation can be snow, sleet or rain and normally we can't see evaporation. Yet in this demonstration, precipitation was only "rain" and we could see the steam.)

Part B: Groundwater and Leachate

Time

20 minutes

Vocabulary

groundwater
hazardous waste
leachate
toxic

Materials (for one group or for one demonstration)

1 funnel that will hold 3 cups
1 clear glass container (funnel must be able to rest on top)
1 cup of gravel
1 cup of sand
1 cup of soil
1 spray bottle filled with water
1 cup of dirty water (a combination of water and sand)
red food coloring

Supplementary Information for Teachers

When it rains or snows some water evaporates; some gets used by plants and animals; some goes into lakes, rivers, and oceans; and some goes into the ground. The water that goes into the ground moves through various layers of soil, through sand and gravel, and becomes a sort of underground stream called groundwater. The gravel and sand clean the water. This experiment shows the process of water moving through soil and sand and coming out clean (free of dirt). However, many hazardous chemicals do not get cleaned out by this process and end up in the ground water. A major concern with landfills is that the leachate (the liquid that drains out of a landfill) will contain toxic materials that cannot be cleaned by the sand and gravel and therefore will end up in the water table. Many wells get water directly from the water table. The red food coloring in this experiment demonstrates how a small drop of hazardous waste can affect a large amount of water.

Activity Procedure

Note that this activity can be done as a demonstration or in small groups

1. The land will be represented by layers of soil, sand, and gravel. To set up your "land," position the funnel so it rests on the container.

2. Put one cup of gravel in the funnel, followed by one cup of sand, followed by one cup of soil. Explain to the students that this represents the land.

3. Saturate the "land" with the spray bottle until water is dripping from the funnel.

4. Discard the first 1/4 cup of water that comes through the funnel and then continue watering.

5. Ask: What happens to the water as it goes through the funnel?

6. Add the cup of dirty water. Ask: Does the dirty water come out cleaner? Why? What might slip through?

7. Add some "hazardous waste" (red food coloring) to the soil and continue watering. Ask: What happens to the water?

8. Conduct a class discussion about how the demonstration relates to leachate from landfills and our drinking water supply using the following questions:

> * How does hazardous waste get into our drinking water? (Hazardous materials can be thrown out with regular trash, put in a landfill, and leached into the water supply. For example, motor oil is a hazardous item commonly thrown out with regular trash.)
> * Why is it illegal to throw hazardous waste in the woods or near a pond?
> * Why are some people so concerned about the hazardous waste in landfills?
> * What would happen if an unlined landfill had toxic chemicals in it?
> * What can be done to minimize leachate from landfills? (source reduction of hazardous materials)
> * Can you think of ways to contain leachate so that it won't seep out? (materials that water won't pass through, such as clay or plastic used as liners)
> * What problems would exist if leachate were contained within the landfill? (It would accumulate. Presently, in lined landfills, leachate is pumped out and disposed of in a hazardous waste disposal site.)

Part C: The Oxygen and Nutrient Cycles

Time

30 minutes

Vocabulary

decay   oxygen cycle
nutrient   photosynthesis
nutrient cycle

Activity Procedure

1. Ask students if they can think of other natural cycles besides the water cycle. To help elicit responses, ask the students where we get our oxygen from. Is it something we create, or can it run out? With students' help, draw the oxygen cycle on the board and explain the cycle. Plants give off oxygen as a waste product of photosynthesis. Animals take in oxygen as they breathe and exhale carbon dioxide. Plants use the carbon dioxide for photosynthesis.

2. Are there any other cycles? What happens to dead leaves and trees when they die? Where does soil come from? With students' help, draw and explain the nutrient cycle. (If students have made compost piles, they should have a good understanding of this cycle.) All living things need food or nutrients to grow. Plants take up nutrients from soil to make sugar. Animals then eat the plants. They return nutrients to the soil through body wastes. When plants and animals die and decay, nutrients are returned to the soil.

3. Discuss with students ways in which humans have either helped or hurt the natural cycles of oxygen and nutrients. This can be done as a class or small group discussion.

THE WATER CYCLE

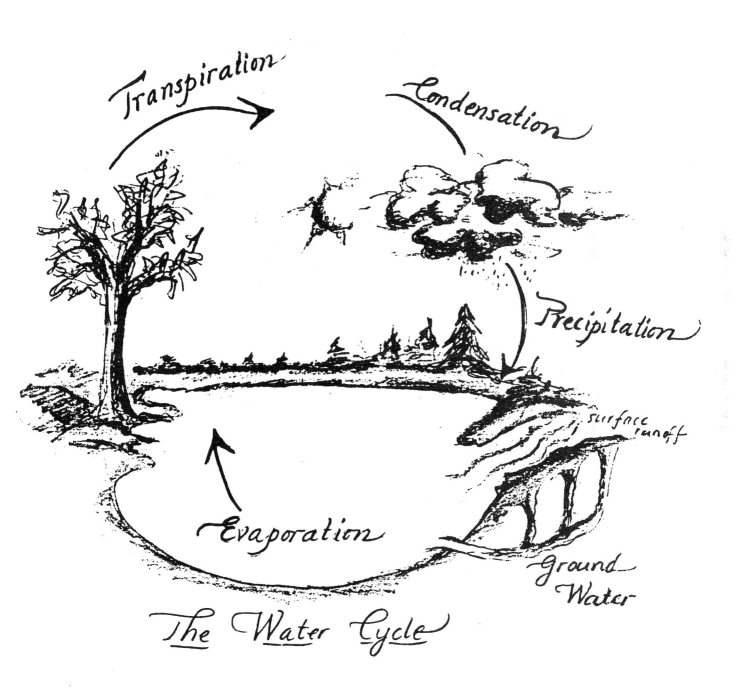

* Reprinted by permission of Joyce Chertow

Activity 8

BURNING CANS

Objective

By creating "burning cans" (cans in which trash is burned), students will demonstrate the effects of incineration on the weight and volume of trash, as well as on the production of emissions and ash.

Note

This activity does not simulate the operation of state-of-the-art incinerators, but it will illustrate reduction of the weight and volume of trash as well as the production of emissions and ash. The U.S. Environmental Protection Agency claims that today's incinerators capture most of the emissions.

Take several precautions. This activity **should be done outside** since it will probably set off fire alarms inside. Whether you do it outside or inside, **check with your school's administration first** about using fire in a school lesson. **Do not burn plastic**, as it can produce toxic emissions. Also, **students should be warned not to try this experiment on their own at home or unsupervised.**

This activity can be done by students in small groups or as a demonstration by the teacher, which would alleviate certain safety concerns. Teachers may also decide not to do the actual burning activity and just hold a discussion. Most students have seen things burn or could use their imaginations to participate in a discussion on the advantages and disadvantages of incineration.

Suggested Previous Activities

1. How Much Trash Do You Throw Away in a Week?
3. Identifying Natural Resources
5. Constructing a Mini-Landfill
7. Natural Cycles

Time

45 minutes

Vocabulary

emission	resource recovery facility
incineration	waste-to-energy facility

Materials (per group of 3 students, or use one larger set as a teacher demonstration)

Handout 11, "Burning Cans Analysis Sheet"
one 1-gallon metal can with several ventilation holes about 1 inch from the bottom
a piece of metal screen large enough to cover the top of the can
5 pieces of cardboard, 6 inches square
masking tape
enough solid waste to fill a box 6" x 6" x 6"
lunch scraps such as orange or banana peels, egg shells
small milk cartons, napkins, newspaper, etc.
matches
ruler
scale (in ounces)

Activity Procedure

1. Explain to students that they will be burning trash in cans to see how incineration effects the amount of trash in weight and volume. Ask students to brainstorm what materials will burn completely and how the amount of trash will change by burning.

2. Distribute student worksheets.

3. Give students the following instructions:

A. Tape cardboard pieces together to form a box with four sides and a bottom. Then fill the box with waste, weigh the box, and calculate the volume as explained on the worksheet.

volume = length x width x height
volume = 6 in x 6 in x 6 in = 216 cubic inches

B. Transfer the waste from the cardboard box to the gallon can.
C. Light the materials on fire and immediately place the screen on top of the can.
D. Observe what comes out of the can while materials are burning and record your observations on the appropriate area on the worksheet.
E. When burning is complete and ashes have cooled, return the ashes to the cardboard box, spreading them evenly on the bottom of the box.
F. Measure the depth (height) of the ash layer.
G. Using the same formula above, calculate the volume of ash.
H. Now calculate the difference in the volume occupied by the waste before and after burning.
I. Weigh the box containing ashes.
J. Calculate the difference in weight before and after burning by subtracting the weight of the box containing ash from the weight of the box containing unburned trash.

4. Conduct a class discussion in which students consider the advantages and disadvantages of incineration based on their observations. These questions may foster discussion:

* What disappeared?
* What kinds of items burn completely?
* What kinds of items don't burn well?

* In what ways is incineration a good method to reduce the amount of trash?
* Is anything given off into the air during incineration? If yes, how do you think this effects the air?
* In what ways is this incinerator similar to or different from a real one? (Real incinerators are usually waste-to-energy facilities, they have scrubbers and anti-pollution devices to help decrease air pollution.)
* What are the advantages of incineration? What are the disadvantages?

BURNING CANS

Name: _____ **Date:** _____

Calculate the volume of the waste to be burned:

Volume = length x width x height.

Multiply: _____ x_____ x _____

Volume of the waste = _____

What is the weight of the waste? _____

What comes out of the can while the waste is burning?
Record your observations here:

Calculate the volume of ash:

Volume = length x width x height

Multiply: _____ x _____ x _____

Volume of the ash = _____

Weight of the ash = _____

How much has incineration reduced the volume and weight of the waste?

Volume of waste _____ minus volume of ash _____ equals

_____ cubic inches or _____ % of the original waste reduced.

Weight of waste _____ minus weight of ash _____ equals

_____ (weight measure) or % of the original waste reduced.

Reprinted by permission of Association of Vermont Recyclers

Activity 9

A BRIEF HISTORY OF TRASH DISPOSAL

Objective

Students will learn about changes in methods of trash disposal over the past 200 years.

Suggested Previous Activities

1. How Much Trash Do You Throw Away in a Week?
2. Changing Times, Changing Waste
7. Natural Cycles

Time

45 minutes for introduction
1 homework assignment (reading)
45 minutes for follow up discussion

Vocabulary

ground water contamination   solid waste management
incineration   trash-to-energy
landfill

Materials

paper and pencils
1 copy of Handout 12 for each student

Activity Procedure

1. Discuss with students the many ways trash is disposed of today. Do they think this is the way trash has always been disposed of? What might have been done in the past?

2. Have students write a short history about the ways they think trash has been disposed of, starting in the 1700's. These questions may help them create their histories:

* What kinds of things were thrown out in the 1700's compared to now?
* What conditions of life in the 1700's may have effected the way trash was disposed of?
* What events or problems may have led to the current ways of dealing with trash? (What happens when you leave trash outside?)

3. Have students share their histories with the class.

4. Have students read Handout 12, "Getting Rid of the Garbage" in class or as homework. Use the similarities and differences between the students' imagined histories and the history recounted in the reading to discuss changes in trash disposal over the years.

5. Additional work on this topic could include having students research the history of trash in their community. This information can be difficult to find, so teachers should check on resources (whether books or local officials) before giving the assignment. Students could investigate these questions:

  *Is trash the responsibility of the individual, town, city, county or state?
  *Is there or was there a local dump or landfill? Has it changed over the years?
  *What laws have been passed that effect what happens to your trash?
  *What local citizens' groups exist that try to help deal with this issue? When and why did they start?

GETTING RID OF THE GARBAGE

From the 1700's

Throughout history, people in the United States seem to have thrown out their trash in the cheapest and quickest ways possible. From the 1700's until the mid-1950's, communities in many states disposed of their solid wastes by burning and dumping them. Trash was removed to places few people had ever lived, such as river banks, wetlands, floodplains, marshes, swamps, and bogs.

By the mid-1880's, these dumps looked and smelled terrible and were causing a number of health problems. They were attracting all types of rodents and other pests that pass along diseases. As populations grew, so too did the amount of garbage, and the problem of what to do with household wastes intensified. By the late 1800s, communities were making laws to clean up these dump sites.

In New York City, during the early 1800s, trash was typically gathered into a pile by residents and placed in the street, where it was supposed to be collected by the city. Most often it was not collected and was left in the streets. Roaming pigs, goats, and dogs became the city's only street cleaners. Trash that was collected was either dumped in empty lots or over a pier into the ocean. Many times the garbage was never carried out with the tide, so it ended up in large piles back at the pier. Then the deadly disease cholera began to spread, in part from all the garbage lying around. The city realized the importance of proper trash disposal in trying to stop the epidemic. Clearly the trash had to be picked up and taken away. But to where?

The Wetlands Are Chosen

At the beginning of this century, most communities in the United States dumped their waste in marshes and wetlands. Many people did not think these lands had any value--you couldn't build a house or a store on them-- so the land could be bought cheaply as a place to dump the trash.

People thought the wet soil would help clean the residues from the trash; they believed the extra water would dilute the trash residue. But something very different happened. Dangerous chemicals from the trash worked their way down into the land and the water below the marshes This is called ground water contamination. Soon the streams and lakes that supplied water for people to drink and use in their homes and businesses were contaminated too.

A Burning Issue

In the 1930's a lot of the garbage was burned in open pits. Burning trash down to ashes made the volume of trash decrease, and that seemed to be a good idea.

Unfortunately, there were just too many fires burning. People who lived near the pits were breathing smoky air. Fire departments always seemed en route to putting out trash fires. In fact, there were so many fires in the open pits that they were used by fire departments to train new fire fighters.

Finally people responded by limiting open burning to very specific areas, and after a while there just wasn't enough land left that could be used for this. Another solution was to build big incinerators so that less land would be needed. We still incinerate some trash today.

Making New Laws

By the 1960's and 1970's, most communities in Massachusetts were still dumping their trash in areas called "sanitary landfills". Many sanitary landfills were still located in or near wetlands and other places that help us get the water we need. People became concerned about contaminating the water.

Many environmental groups, such as the National Audubon Society and the Sierra Club, led the effort to make changes in how we dumped trash. They helped convince the federal government to pass a law in 1965. Called the National Solid Wastes Act, this law required every state in the United States to dispose of its trash in more careful ways.

Massachusetts then made its own laws that were even stricter. There could no longer be open garbage dumps. All new sanitary landfills had to be built to protect the health and safety of people and the environment. Old dumping grounds had to be cleaned up. Approximately thirty towns closed their landfills. Many others made improvements in the landfills they had. Some towns could do nothing, and the old methods of dumping continued because they didn't have any other places to bring their trash.

The Environmental Protection Agency (EPA) was created in 1970 as a result of many citizens wanting the government to make the air and water cleaner. For the first time the federal government could fine companies that would not stop polluting. Companies can pollute less by changing how they operate and by installing anti-pollution equipment. However, it can cost a lot of money to get the equipment needed to reduce air and water pollution. In Massachusetts, only two incinerators could afford to stay open, and many landfills had to close. In 1975 the first successful trash-to-energy plant in the world opened on the site of an old landfill.

Recent Changes

One law that has affected everyone in Massachusetts and several other states has been the Bottle Bill. Starting in 1983, a deposit was required on the bottles and containers used to hold many types of drinks. When people return a used bottle to the store or collection center, they get back money and help reduce the amount of trash. The original purpose of the Bottle Bill was to reduce litter and conserve raw materials, but it also helped many people realize how much trash they had been throwing out every day. Now many people are recycling other items, such as newspapers and cans.

Changing the ways towns dispose of trash can be very expensive. In 1987 the state of Massachusetts established the Solid Waste Act. This act set aside a lot of money to help towns dispose of their trash in safer ways. The state hopes people will build new and better solid waste disposal facilities. Many towns now require their citizens to separate bottles, cans, and newspapers from the rest of their garbage, and a truck comes around and picks up these items to be recycled. This is one way of reducing the amount of land used for dumping garbage.

Sources used to write the brief history of trash disposal:

The Solid Waste Management Resource Guide for Massachusetts Schools

The Cholera Years, Charles E. Rosenberg

Activity 10

DO YOU KNOW WHERE YOUR TRASH IS?

Objective

Students will learn what happens to trash in communities near them by interviewing town officials and charting their findings.

Note

This activity requires students to do telephone research. You may want to contact the people your students will be calling and ask their permission to have students call them.

Telephone research can sometimes be intimidating or frustrating for students. If students role-play possible telephone conversations and practice good telephone etiquette before they make their calls, they will probably have more success and feel more confident. Allowing them to work in pairs to support each other in telephoning is sometimes helpful. Also, it is important to warn students that they will probably be passed from one official or department to another before they actually find the information they need. They'll need to be persistent. Students should not leave the research for the last minute, since it is sometimes difficult to reach people.

There may be some logistical challenges with this activity. It can be difficult for students to get to a phone, and many students may end up calling the same town official at the same time. To solve these problems try doing this activity as a voluntary extra-credit activity or as a class project with a speaker phone.

This activity may be difficult for students whose families live near a landfill or incinerator. They may feel shame or be subject to comments by students who live in other parts of the community. The activity may highlight the differences in economic classes and family choices.

Suggested Previous Activities

1. How Much Trash Do You Throw Away in a Week?
5. Constructing a Mini-Landfill
8. Burning Cans

Time

20 minutes to set up the assignment
week of homework (data collection)
45 minutes once data is assembled

Vocabulary

incinerator   tipping fee
landfill   waste-to-energy facility
recycling

Materials

1 copy of Handout 13, "Where Trash Goes" for each student or pair of students (If
     a student is researching more than one town, she or he will need extra
     copies of the handout--one per town researched.)
large piece of mural paper (4' x 6') to draw map on
a map of the town and surrounding communities (To make a large map for
     students to trace trash disposal, either sketch the map onto large paper, or
     make a transparency of the map and project it onto the large paper and
     copy the outline.)
markers (3 different colors)
colored yarn (2 colors to match two of the markers)
tape
2 pieces of poster paper
1 copy of Handout 14, "Taking Out, and In, the Garbage," for each student

Activity

1. Explain to students that they will phone officials in surrounding communities to find
out how the town's trash is disposed of. Tell them the best place to start is the waste
management division of the public works department at the town or city hall. Teachers
may want to check first and find the accurate department name for each community
involved.

2. Go over telephone etiquette with them, perhaps having a student play herself or
himself and another play an official. Callers should always explain who they are and why
they are calling.

3. Students make their calls during the next few days, recording the information on
Handout 13, "Where Trash Goes."

4. At the end of the week, students should bring worksheets to class.

5. Post the large map of towns and surrounding communities on the wall so that students
can draw on it.

6. Make a key for the map, using one color marker to indicate landfilling, one for
incineration or waste-to-energy facilities, and one for recycling. The colors used for
landfilling and burning should correspond with the colors of the yarn.

7. Students whose towns use landfills should color their towns with the appropriate
marker. Those whose towns use incineration or waste-to-energy facilities should also
color their towns according to the key.

8. To indicate recycling, students can use the appropriate markers to make recycling
arrows on top of the other background color for landfills or incinerators.

9. If a town transports its trash to be landfilled or burned, students should color the town with the appropriate marker (landfilling or incineration) and tape a piece of yarn with the matching color to connect the town with the location of the landfill or incinerator.

10. Tipping fees can be placed on a chart or bar graph on poster paper indicating the towns along the horizontal axis, and tipping fees along the vertical axis.

11. Recycling percentages can be placed on a chart or bar graph on poster paper indicating towns along the bottom and percentages along the left margin.

12. Hold a class discussion with students on their thoughts and conclusions using the following questions:

> * Ask students to study the map after their results have all been recorded and colored in. What do they notice about how trash is disposed of in their communities?
> * Do more towns use landfills, waste-to-energy facilities, or incinerators? Why might one be more common than the others in your area?
> * Is there a correlation between the distance trash is transported and the tipping fee?
> * Is there any correlation between the method of trash disposal and the tipping fee?
> * Were you surprised by what you learned? If so, what surprised you?
> * What factors might go into a decision about which form of disposal to use? (population density, geography, economics)
> * In your town, would you prefer an incinerator, a waste-to-energy facility, or a landfill? Why?
> * Are there some types of areas in which one facility would be more appropriate than the others?
> * What might be the differences between how garbage is diposed of in urban versus rural areas?

13. Pass out Handout 14, "Taking Out, and In, the Garbage." Use the following questions to foster a discussion:

> * Does anything on this chart surprise you? Explain.
> * Are there any states that don't export? Are there any that don't import?
> * What further questions does this information raise in your mind?

WHERE TRASH GOES

1. What is the name of the town you researched?

2. What is the name and title of the person you spoke to?

3. Explain how the town's trash is disposed of. **Be specific.** (If the trash is landfilled, where is the landfill located? If it is incinerated or burned in a waste-to-energy facility, where is the facility located? What happens to the ash after incineration?)

4. Is anything recycled? **Be specific** about what is recycled.

5. What percent of the town's trash is recycled?

6. What is the tipping fee?

7. Is the town facing any problems or changes in relation to waste disposal?

Taking out, and in, the garbage
Interstate movements of municipal solid waste.

NY Times Sept. 8, 1991

STATE	IMPORTS GARBAGE FROM	EXPORTS GARBAGE TO
Alabama	Georgia, Tennessee	Florida, Georgia, Mississippi
Arizona	California	California, Nevada, Utah
Arkansas	Oklahoma, Texas	Louisiana, Mississippi, Missouri, Oklahoma, Tennessee, Texas
California	Arizona	Arizona, Nevada, Utah
Colorado	Nebraska, New Mexico	Nebraska
Connecticut	Massachusetts, New York, Rhode Island	Massachusetts, Ohio
Delaware		Pennsylvania
District of Columbia		Maryland, North Carolina, Pennsylvania, Virginia
Florida	Alabama, Georgia, New York	Georgia
Georgia	Alabama, Florida, South Carolina, Tennessee	Alabama, Florida, South Carolina, Tennessee
Idaho		Washington
Illinois	Indiana, Iowa, Michigan, Minnesota, Missouri, New Jersey, New York, Wisconsin	Indiana, Kentucky, Michigan, Missouri, Wisconsin
Indiana	Illinois, Kentucky, Michigan, New Jersey, New York, Ohio, Pennsylvania	Illinois, Kentucky, Michigan, Ohio
Iowa	Wisconsin	Illinois, Minnesota, Nebraska
Kansas	Missouri, Oklahoma	Missouri
Kentucky	Illinois, Indiana, New Jersey, New York, Ohio, Pennsylvania, Tennessee, West Virginia	Indiana, Ohio, Tennessee, West Virginia
Louisiana	Arkansas, Texas	Mississippi, Texas
Maine	Massachusetts, New Hampshire	Massachusetts, New Hampshire
Maryland	District of Columbia, New Jersey, New York	Pennsylvania
Massachusetts	Connecticut, Maine, New Hampshire, New York, Rhode Island, Vermont	Connecticut, Maine, New Hampshire, Vermont
Michigan	Illinois, Indiana, New Jersey, Ohio, Pennsylvania	Illinois, Indiana, Ohio
Minnesota	Iowa, Wisconsin	Illinois, North Dakota, South Dakota
Mississippi	Alabama, Arkansas, Louisiana, Tennessee	Tennessee
Missouri	Arkansas, Illinois, Kansas, New Jersey, New York	Illinois, Kansas
Nebraska	Colorado, Iowa	Colorado, South Dakota, Wyoming
Nevada	Arizona, California	
New Hampshire	Maine, Massachusetts, Rhode Island, Vermont	Maine, Massachusetts, Vermont
New Jersey		Illinois, Indiana, Kentucky, Maryland, Michigan, Missouri, Ohio, Pennsylvania, Virginia, West Virginia
New Mexico	Texas	Colorado, Texas
New York		Connecticut, Florida, Illinois, Indiana, Kentucky, Maryland, Massachusetts, Missouri, Ohio, Pennsylvania, Vermont, Virginia, West Virginia
North Carolina	District of Columbia, South Carolina	South Carolina
North Dakota	Minnesota	South Dakota
Ohio	Connecticut, Indiana, Kentucky, Michigan, New Jersey, New York, Pennsylvania, West Virginia	Indiana, Kentucky, Michigan, Pennsylvania, West Virginia
Oklahoma	Arkansas, Texas	Arkansas, Kansas
Oregon	Washington	Washington
Pennsylvania	District of Columbia, Maryland, New Jersey, New York, Ohio	Indiana, Kentucky, Michigan, Ohio, Virginia, West Virginia
Rhode Island		Connecticut, Massachusetts, New Hampshire
South Carolina	Georgia, North Carolina, Virginia	Georgia, North Carolina
South Dakota	Minnesota, Nebraska, North Dakota	
Tennessee	Arkansas, Georgia, Kentucky, Mississippi	Alabama, Georgia, Kentucky, Mississippi
Texas	Arkansas, Louisiana, New Mexico	Arkansas, Louisiana, New Mexico, Oklahoma
Utah	Arizona, California	
Vermont	Massachusetts, New Hampshire, New York	Massachusetts, New Hampshire
Virginia	District of Columbia, New Jersey, New York, Pennsylvania	South Carolina
Washington	Idaho, Oregon	Oregon
West Virginia	Kentucky, New Jersey, New York, Ohio, Pennsylvania	Kentucky, Ohio
Wisconsin	Illinois	Illinois, Iowa, Minnesota
Wyoming	Nebraska	

Activity 11

WHAT SHOULD BE DONE WITH ALL THE TRASH?

Objective

Students will explain the ways in which we dispose of garbage and describe some of the controversies associated with the disposal methods.

Suggested Previous Activities

5. Constructing a Mini-Landfill
7. Natural Cycles
8. Burning Cans
10. Do You Know Where Your Trash Is?

Time

1 hour homework assignment (reading)
45 minutes for follow-up discussion

Vocabulary

incineration   recycling
integrated waste management   resource recovery
landfill   source reduction
leachate   waste-to-energy

Materials

1 copy of Handout 15, "Fill 'Er Up,"or handout 16, "Garbage, Garbage
        Everywhere" for each student. Either article can be used, depending on
        students' reading ability. "Fill 'Er Up" is the easier of the two articles.
1 copy of Handout 17, "Trash and Waste-to-Energy Facilities in Centerville" for
        each student
paper and pencils
1 copy of Handout 18,"Disposal Methods" for each group of students
poster materials for each group: markers (non-toxic) and poster paper (recyclable)

Supplementary Information for Teachers

The definitions of incineration, waste-to-energy, and resource recovery facilities can be confusing. Essentially, waste-to-energy and resource recovery facilities are the same. In both methods, trash is burned and energy from the burning is used to generate electricity. Incinerators simply burn trash. Incineration is an older method and is now being replaced by waste-to-energy and resource recovery facilities.

Activity procedure

1. For homework, have students read the article and the Centerville story before beginning the following activity.

2. Have students work in small groups of three or four to answer the questions on Handout 18, "Disposal Methods." Students should refer to articles and Centerville story to answer questions.

3. Have each group make two posters that might be presented to a town which is considering allowing the construction of a waste-to-energy facility. One poster can represent reasons for opposing incineration and promote other methods of trash disposal and reduction. The other poster can promote the use of incinerators as part of an integrated waste management system.

4. Have each group present their posters to the class and discuss which are most convincing and why.

Alternative Activity Procedure

1. After students have read all the articles for homework and looked at the Centerville story in class, have them write down the questions they have about the information.

2. In class discussion, see how many questions other students can answer from the readings or other information.

3. Have one student keep track of questions that no one can answer.

4. Have students make a list of their concerns about managing solid waste. (These concerns will likely include questions about safety of landfills and waste-to-energy facilities.)

5. For homework, divide the lists of questions and concerns and have students try to find answers through library research or contacting state environmental agencies, local environmental groups, incinerator companies, trash collection companies, local public works disposal departments, and people with different perspectives. on the issue.

6. You or the students may want to invite speakers from the various groups to answer the class's questions in person.

Follow-up Activities

Students can interview relatives and members of the community to find out how they would react to having a landfill or waste-to-energy facility in their town. If the town already has a landfill or waste-to-energy facility, how do they feel about it? Do they have any concerns?

Display the posters in the school to promote the education of other students about trash disposal.

Interview other students in the school to find out their views on trash disposal methods. Display the results along with the posters.

FILL 'ER UP

by Billy Goodman

Let's talk trash. Say, about four pounds a day if you're an average American. If you live in a big city, better make it about six or seven pounds. That's how much trash each American throws out every day. That adds up to more than 150 million tons of household waste a year (call it what you will—trash, solid waste—it's all garbage). New York City residents throw out 27,000 tons *a day*.

Where does it all go? Most—90 percent—winds up in *landfills* (dumps). The rest is *incinerated* (burned) or recycled. As landfills fill up, these other options are getting renewed attention.

BURY IT

Not far from the back yards of some people on New York's Staten Island is the world's largest landfill, Fresh Kills (named for the two creeks, or *kills*, that it borders). Fresh Kills' four dumping areas accept 25,000 tons of trash a day. By the end of the century, Fresh Kills will be a mountain of trash 500 feet high, the highest point on the East

Coast. Then there will be no more room at the dump. Other landfills around the country are running out of room even faster.

Meanwhile, six days a week at Fresh Kills, 24 hours a day, cranes unload barges, and trucks carry their loads to the growing mountain.

Last winter I visited Fresh Kills with Henry Perahia, Director of Engineering for New York City's Department of Sanitation. As we drove around the dump, Perahia taught me some of the finer points of garbage. How, from far away, you can locate the *active bank* (the small area where garbage is being dumped and spread). "You look for the sea gulls," he says. "They dine from the finest restaurants in the world." How the Department of Sanitation is planning to use a deodorant spray on the garbage this summer. How specially designed boats cruise the water around the barges to scoop up trash.

We watched as a crane took 5,000-pound gulps of garbage from a barge and spit each mouthful into a tracked wagon. When two wagons were full, a tractor slowly pulled them—on a road made of garbage—to the active bank, where they dumped their load

The vehicles pictured in the landfill (*below*) have special tires that help shred and compact the garbage. In a typical incinerator (*bottom*), trucks dump garbage into a pit at one end of an enormous hangar. Cranes drop the garbage into a furnace. The superheated gases boil water to make steam and, ultimately, electricity. Material that doesn't burn, ash, is removed at the bottom and after the gases go through the pollution-control equipment. The ash contains hazardous materials.

Courtesy of Wheelabrator Environmental Systems

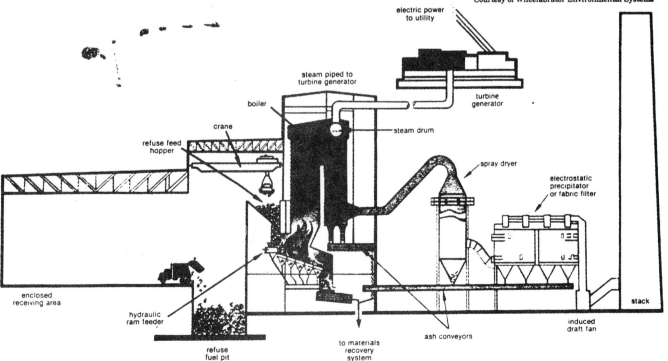

without even stopping.

Landfills aren't very pretty to look at, and many pollute their environment, too. Water percolates through garbage like water through grounds in a coffee pot. But instead of coffee, you may end up with a toxic brew that can contaminate underground wells tapped for drinking water. State-of-the-art landfills have clay and plastic liners to contain the *leachate* (liquid containing dissolved material) and pipes to collect it. Yet all the safety features in the world can't make more room for landfills. Some cities and towns are planning to incinerate more of their trash.

BURN IT

This must be what hell is like, I thought, looking through a small, thick window at a raging inferno. The heat was so intense, I couldn't look inside for very long. I was standing outside the furnace of an incinerator in Peekskill, New York. There, on the banks of the Hudson River, three 100-foot-high boilers burn 2,000 tons of trash a day.

I bet I know what you're thinking: There's the solution to the garbage crisis. Just burn it all and we won't have to worry about running out of room at the landfills. That strategy sounds even better when you learn that the trash in the Peekskill incinerator stokes a 2,500°F fire that creates steam that drives a turbine that makes electricity. About 55 megawatts. Enough for 55,000 people, or 18 suburban shopping malls.

But there's a catch. Not all the garbage burns, even at 2,500°F. Paper, plastic, and wood vaporize, but leave behind heavy metals, like lead, cadmium, and mercury. The metals stick to particles of other unburned material known as *ash*. At Peekskill, the 700 tons of ash left after a day's burning is hauled to a special landfill. Rainwater seeping through the landfill might remove the metals from the ash and contaminate groundwater.

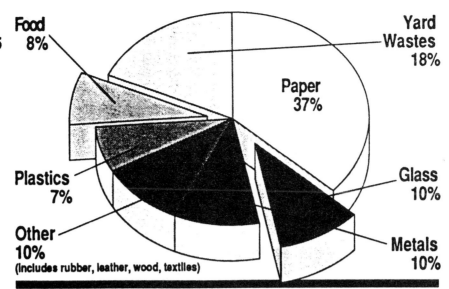

Food 8%

Yard Wastes 18%

Paper 37%

Plastics 7%

Glass 10%

Other 10% (Includes rubber, leather, wood, textiles)

Metals 10%

The pie chart (*above*) shows what's in the solid-waste stream generated by communities across the nation. As shown by the recycling of aluminum cans and newspapers (*left*), much of the material that we now throw out or burn could be recovered for its raw materials. Source: Franklin Associates.

As if that's not enough, it turns out that better air-pollution-control equipment creates more small ash particles. "You're just trading one pollution problem for another," says Michael Herz, an attorney with the Environmental Defense Fund. EDF wants ash treated as hazardous waste. With many other groups, EDF favors recycling much more of what we now throw away.

RECYCLE IT

Here's a riddle for you: I walked up to a vending machine. I put in something metallic. I got something metallic in return. What did I do?

No, I didn't buy a can of soda pop. I bought a nickle with a can. Several states require a deposit on bottles and cans to encourage recycling. When you return the bottle or can, you get your deposit back.

Only five percent of solid waste is recycled nationwide (much more in some states such as Washington and New Jersey). In order for that figure to grow, says Wayne DeFeo, recycling coordinator of New Jersey's Somerset County, attitudes must change.

"We don't pick up garbage," he says. "We don't pick up trash. We pick up resources." Somerset County recycled 15 percent of its solid waste last year. The other 85 percent? Trucked to a Pennsylvania landfill at a cost of $100 a ton.

What can be recycled?
● **Aluminum:** Each pop can recycled saves six ounces of gasoline. That's how much energy is required to make a new can from scratch.
● **Glass:** Bottles and jars can be made into more of the same, added to asphalt, or used in insulation.
● **Paper:** All kinds of paper can be remade into stationery, packaging, paper towels. Many newspapers are printed on recycled paper.
● **Plastic:** Yes, even some kinds of plastic are recyclable. Plastic soda bottles made of polyethylene terephalate (PET) can be processed into a geotextile, a feltlike material that has protected the turf at Michael Jackson and Prince stadium concerts.

The garbage problem isn't about to go away. But we can make it less painful by creating less trash to begin with, by using durable products instead of disposable ones, and by reusing some of what we throw away. And we can insist on environmentally safe landfills and incinerators for the real garbage we create. If there's less garbage, only the sea gulls will complain.

YOU BE THE SCIENTIST
1. What are the most common materials in garbage? Which ones are the easiest to recycle?
2. List some advantages and disadvantages of each way to handle trash (landfilling, incineration, recycling).

Environment

Garbage, Garbage, Everywhere

Landfills are overflowing, but alternatives are few

Leading a climbing team up Mount McKinley, the highest peak in North America, Captain Richard Garrison, an Army chaplain, discovered that even the remote Alaskan wilderness has been despoiled. There, at 8,500 ft., was a pile of garbage—partly eaten food, foil wrappers from freeze-dried meals, plastic bags and other trash left behind by previous climbers who had disobeyed the basic outdoor rule to backpack out all such junk. "It really detracts from the experience," says Garrison.

But at least the garbage was out of the sight and smell of most Alaskans. Some Chicagoans are not so fortunate. "As you bike past certain streets, you are overwhelmed by rancid smells of rotting garbage," says Jeannie Little of Greenville, S.C., passing though the Windy City on a tour of the U.S. From an apartment in a pricey neighborhood she can see rats in the alley below snacking on spilled morsels. Says Little: "I'm horrified by the fact that we generate so much garbage and don't have a place to put it."

That is a simple but accurate description of a situation approaching the crisis stage throughout the U.S. The affluent, fast-paced, throwaway American culture is producing trash on a stupendous scale. Between 1960 and 1986, the amount of American garbage grew 80%, from 87.5 million tons to 157.7 million tons annually. It is expected to increase 22% by the year 2000, when the malodorous mound will weigh 192.7 million tons.

In a study comparing the U.S. with other nations, its pile of disposable diapers, melon rinds, grass clippings, plastic hamburger boxes, broken mattresses and discarded tires came to 1,547 lbs. for every man, woman and child in the country. Only Australians came close to producing as much waste: a prodigious 1,498 lbs. per person. The average West German or Japanese threw away about half as much. But even the U.S. figure pales next to that of California, where some calculations have the average citizen throwing away 2,555 lbs. a year. Says Attorney Jill Ratner, who is active in environmental causes: "In Los Angeles County we generate enough trash to fill Dodger Stadium with garbage every nine days or so."

Dodger Stadium, however, is not available for the purpose, and in a growing number of communities, neither is anything else. About 80% of U.S. trash is disposed of by burying it under thin layers of earth at a site known as a landfill. But an estimated half of the landfills in the country have filled and closed in the past decade, leaving about 9,200 with space remaining. Some 6,000 belong to counties, cities and towns. The Environmental Protection Agency projects that one-third of these will run out of space and shut down in the next five to ten years. In some areas the day of garbage reckoning is much closer. The two landfills in San Mateo County, near San Francisco, will be full next year, the one in Omaha's Douglas County in no more than 15 months.

Opening a new landfill, at least anywhere near a big city where they are most needed, is next to impossible. The NIMBY (not in my backyard) syndrome takes over. In this case, public fear is to some extent justified. In addition to being unsightly and smelly, all too many landfills leak pollutants (discarded cleaning fluids, for example) into groundwater.

The EPA last week moved to stop such pollution, and in the process may have hastened the day when there will be no place to dump garbage. The agency proposed regulations that would force all municipal landfills to monitor hazardous wastes and methane gas; ban discharge of harmful wastes into underground water; and strengthen controls on rodents, insects, fire and odor. While the new standards are reasonable—some environmentalists consider them too lenient—EPA Assistant Administrator J. Winston Porter says "very few" of the 6,000 municipal dumps could meet them now. When the limitations take full effect in 1991, some landfills may be forced to close while some new ones will probably not open.

The EPA estimates that its landfill rules will eventually add $800 million to $900 million a year to the nation's garbage-disposal costs, now calculated at $4 billion to $5 billion annually. Virtually all of the additional load will fall on states and localities, intensifying what for many is already a heavy financial burden. As landfills have become choked and disposal more difficult, "tipping" fees paid by cities to contractors to dump their trash have rocketed from $10 a ton a decade ago to more than $100 in some sections of the Northeast. In the Midwest, Leroy Bannister, administrative assistant to Chicago Mayor Eugene Sawyer, reports, "In 1987 the city had a disposal budget of $19 million. For 1989 it will be $52 million"—and that does not include any spending to comply with the EPA landfill regulations.

What to do? States and cities are pursuing a number of strategies to get rid of their garbage:

SHIP IT. The garbage crisis primarily afflicts the two coasts and major inland metropolitan areas such as Chicago and St.

Louis. Large areas of the country still have open land for landfills, and garbage brokers are scouting those areas for potential trash dumps. But the South is mobilizing to stop any invasion of Yankee garbage. Environmental officials of eight Southern states conferred informally last spring to figure out a legal way to discourage garbage imports. Because of a Supreme Court decision ruling that garbage is a legitimate item of interstate commerce, out-of-state trash cannot be banned outright. But the Georgia general assembly in March passed a law making garbage imports difficult. Third World countries too are rebelling against garbage dumping by industrialized nations. Nigeria is putting a score or more of its citizens on trial for facilitating the dumping of dangerous Italian wastes; if convicted, they could be shot.

BURN IT. Incinerators can reduce garbage weight 70%, as well as produce heat that can be sold to generate electricity, thus reducing garbage-disposal costs. In addition, incinerator builders assert that new technology has removed many of the pollutants that escaped from chimneys in the 1960s and 1970s. So it is not surprising that a new generation of incinerators is developing. But NIMBY operates here too: some Minneapolis residents have mounted strong protests against a burner under construction near the downtown area. Like anti-landfill sentiment, opposition to incinerators has a reasoned basis. Environmentalist Barry Commoner insists that incinerators actually synthesize dioxin, a highly poisonous substance. True, scrubbers and other filters can eliminate dioxin from smoke, but not its concentrated form in the ash residue, causing a prickly problem of how and where to get rid of that hazardous waste.

RECYCLE IT. Many communities are taking a new look at this clean and, in some ways, most efficient solution. Illinois Governor James Thompson last week signed into law a bill requiring 18 of the state's largest counties, as well as metropolitan Chicago, to develop by March 1991 comprehensive waste-management programs that emphasize recycling. Said the Governor: "We're simply running out of room, out of time and out of money for facing these [garbage-disposal] problems in the same old way." EPA Administrator Porter has set a goal of having 25% of U.S. garbage recycled by 1992, vs. 10% now. Still, he concedes, recycling success will only delay rather than avert the day when landfills cannot take any more trash. Main problem with recycling: many Americans simply refuse to be bothered with sorting and separating garbage into recyclable and non-recyclable parts. Nor is there any practical way to compel them to do so.

Technology may help ease the looming crisis. One of the most troublesome elements in the garbage stream is the soaring use of plastic, which is difficult to burn or recycle and, because it is not biodegradable, will clog landfills for centuries. Early efforts to produce plastics that decay were less than successful: some disintegrated under sunlight, unavailable at the bottom of landfills. Others came apart after contact with water, causing supermarket executives to shudder at the thought of what would happen to the groceries in a plastic shopping bag containing a leaky milk bottle. But now there is a method of adding cornstarch to some plastics. Bacteria eat the starch, causing the plastic to fall apart into pieces that can be ingested by microorganisms. Result: the plastic disintegrates in four to seven years.

But in large part, the garbage crisis is a cultural crisis. The development of a throwaway, convenience culture helped create this mess; a real solution may require cultural change. For example, more than 20% of U.S. garbage comprises grass clippings and leaves stuffed into plastic bags and left for collection. Householders should simply leave that grass on their lawns or rake it into a mulch pile, ignoring and thus revising the cultural demand for a golf green-neat lawn. Another cultural change would be required to get Americans to recycle 50% of their trash, as Japanese do. Cultural change is notoriously slow, but it might be speeded up in this instance by the lash of crisis. Americans have always treated garbage as something to be forgotten about the moment it is picked up from the curb. But the day may soon be coming when it will no longer be picked up because there will be no place to take it.
—*By George J. Church. Reported by Steven Holmes/Washington and Elizabeth Taylor/Chicago, with other bureaus*

Each American generates 25 lbs. of trash a week, according to an EPA study

Breakdown by weight of garbage for an average Tucson household

Metals (cans, foil, etc.) 5.3%

Wood, fabric 4.1%

Glass 7.9%

Plastic 7.3%

Diapers 3.6%

Food 17%

Other 8%

Yard wastes 17.1%

Paper (packaging, cardboard, etc.) 18.6%

Magazines, newspapers 11.1%

*Includes dirt and unrecognizable materials
Source: The Garbage Project, University of Arizona

TIME, SEPTEMBER 5, 1988

TRASH AND WASTE-TO-ENERGY FACILITIES IN CENTERVILLE*

An Urgent Call from the Mayor

It's Wednesday afternoon at the mayor's office in Centerville. Maria and Alex, who are aides in City Hall, are called in to see the mayor. She needs information fast because she is going on television next Monday to be interviewed about the problems with trash in Centerville. The mayor has to decide whether she should support the building of a new waste-to-energy facility in the city or propose an alternative method to deal with trash disposal. She tells Maria and Alex to do research for her and report back by Friday afternoon.

"That's just two days from now," Alex says as they leave the office.

"I have an idea," Maria says. "Let's split up. I'll talk to the people who are planning on building the new waste-to-energy facility."

"Yes," Alex agrees. "And I'll speak to the local environmental group to see if they have a recommendation for dealing with trash. Let's meet back here on Friday and share what we have learned."

"Okay. Then we'll be able to give the mayor good advice."

Maria's Report

It is now Friday. Maria and Alex are having lunch, just an hour before the mayor wants their report.

"So what did you find out?" Alex asks Maria.

"I learned so much," Maria says. "The idea of a waste-to-energy facility isn't just to burn trash. As the trash is burned in a high temperature furnace the heat of the fire is converted to electricity, which is very important. Once the trash is burned, it is practically all gone--almost 90% of it. There's just some ash left over, and that can safely be sent to a landfill."

"Do waste-to-energy facilities really make a lot of electricity?" Alex asked. He doesn't sound convinced yet.

"This one company says a typical modern waste-to-energy facility can make enough electricity for 30,000 to 40,000 homes. That's almost every home in Centerville! And you know how expensive it is for Centerville to buy electricity now from a company. Also, don't forget all the problems we have in the summer when everyone is using their air conditioners and we have brown-outs. With a waste-to-energy facility it won't matter how much trash we use, because more trash means more electricity when it gets burned!" Maria thought it was a simple idea.

* This story is based on responses to a series of questions sent to Greenpeace and Wheelabrator Technologies Inc. We appreciate the effort that representatives of both organizations made to help us.

Handout 17 continued

"Is it really safe to burn trash? Won't it be adding dangerous chemicals to our air?" Alex asked.

"It's safer than burning coal or oil, Alex . Sure, there's some pollution, but the company says it is very little. In fact, each waste-to-energy facility they make is safer than the one before it. They say they can make them emit less air pollution than the government standards require.

Alex needed more information. "What about the ash?" he asked. "I understand it can be very toxic."

"The ash is handled very carefully and safely," Maria answered, "and it goes to a landfill. It may be possible in the future to recycle the ash they produce to make asphalt for our new highways!"

"It must be very expensive to build and run one of these waste-to-energy facilities," Alex suggested.

"It is expensive," Maria agreed. "But since a product is being made-- electricity--the cost is actually about the same as other disposal methods."

"But what about recycling?" Alex wondered aloud. "If the company needs our trash people may give up on recycling."

"This is the best part, Alex--the company wants us to recycle! By getting rid of a lot of the glass, aluminum, and paper first, the waste-to-energy facility can do a better job of burning up trash. I think it's a great idea. Let's tell the mayor to support the waste-to-energy facility building."

Alex's Report

"Not so fast," Alex said. "I learned about a lot of problems with waste-to-energy facilities."

"Such as?" Maria asked.

"For instance, can the waste-to-energy facility work perfectly all the time to keep pollution out of the air? Just because you can't see something doesn't mean it doesn't exist. What if the waste-to-energy facility can't catch all the toxins it burns up and slowly leaks out terrible pollutants, and no one can see them! Think of it: each day, day after day, a tiny bit of pollution goes into the air, and it builds and builds. Who would know?"

"What sort of pollutants?" Maria asked.

"The pollutants in burning trash are really awful; dioxins and mercury are the two worst. They're killers! And even the ash is filled with pollutants. Just because the ash goes to a landfill doesn't mean it won't leak into the ground! The environmental group I spoke with is really concerned about the levels of poisons in the environment through air pollution and ash disposal. The danger is an increase in cancer and other health problems. The pollutants add chemicals to the air that contribute to the greenhouse effect, acid rain, and smog. Waste-to-energy facilities don't work well enough to keep the air perfectly clean."

Maria listened silently for a moment. "That's interesting, because the company that builds the waste-to-energy facilities are very proud of the job they do keeping the air clean. They say they're getting better and better at it. What else do we have to think about?" she asked.

Handout 17 continued

"Listen to this," Alex went on. "Almost all the landfills and waste-to-energy facilities are built in poor parts of cities--usually where minority families live. It just seems racist. There are a lot of poor and minority people in Centerville to consider. Where would this facility go in Centerville? Who will live nearest to it?"

"So," Maria wondered, "what do the environmental people think we should do instead?"

Alex nodded his head. "The group was very clear on this point: there is no environmentally safe way to dispose of anything that can't biodegrade. These waste-to-energy facilities make us think that if something is out of sight, then we don't have to think about it. This is very dangerous because we cannot see many of the toxins in our air and land. The facility may decrease the amount of trash but not the amount of pollution."

"You didn't answer my question," Maria said. "I get the idea that there are problems with a waste-to-energy facility. But what should we do instead?"

"The way to deal with the trash crisis is to reduce trash from the very beginning, such as in the factories that make and package what we need. We have to tell the industries that make toxic wastes to find other ways to make their products. If they have to, they can do it. And we can stop buying and using products that can't be reused or recycled. People can be encouraged to buy reusable products, like cloth shopping bags, and to not buy throwaway materials-- styrofoam cups are an example of something we can do without. All waste must be eliminated. The small amount that can't be recycled, reused, or reduced at the source should be landfilled as a last resort. There's no safer or better way to go in the long run, and the long run starts now. At least that's what I was told."

Time's Up

Suddenly Maria looked at the clock on the wall. It was time to see the mayor.

"What should we recommend?" she asked. "She's going to want advice!"

"Right, and it's not only the mayor. All the citizens need to know the facts so they can decide, too."

"This is very difficult," Maria thought as they walked down the hall to the mayor's office. "Centerville has a problem getting rid of trash, and it sure could use electricity from a waste-to-energy facility...."

Alex was thinking at the same time, "Can we afford the risk to our health and the environment of a waste-to-energy facility? I sure wouldn't want one in my backyard...."

They reached the door to the mayor's office and bumped into each other. "You first," they both said at the same time.

DISPOSAL METHODS

1. What are some of the problems of using landfills to dispose of waste? List at least three and explain each.

2. What are some positive aspects of using landfills for disposal?

3. What are some problems with waste-to-energy facilities? List at least three and explain each.

4. What are some of the positive aspects of waste-to-energy facilities?

5. What is source reduction?

6. Do you think source reduction is effective in reducing the amount of trash? Explain.

7. Which disposal method, waste-to-energy facilities or landfills, is better? Explain.

Activity 12

RECYCLING: MAKING PAPER

Objective

By making and selling recycled paper, students will learn how paper recycling is accomplished, while considering how it saves natural resources and reduces the amount of waste. They will also learn about the economics of recycling.

Suggested Previous Activities

10. Do You Know Where Your Trash Is?
11. What Should Be Done with All the Trash?

Time

60 minutes to make paper
2 homework assignments (selling paper and reading)
45 minutes for discussion

Vocabulary

cellulose	recycling
lignin	reusing
millbroke	sludge
post-consumer waste	slurry
recyclable	virgin paper
recycled	

Materials

4-5 dishpans
old paper (anything except newspaper) collected by students
water, several buckets
6 screens (These are very easy to make. Cut pieces of wood to make 5" x 7" frames. Nail frames together. Cut nylon screening to fit over frame. Use staple gun to staple screen onto wood.)
2 blenders
paper towels
newspapers, enough to cover tables
sponges
rolling pin
plant material, such as a leaf or flower (optional)
1 copy of Handout 19, "Recycled Paper - What Is the Story?," for each student

Supplementary Information for Teachers

We have included some information about how the production of virgin paper versus recycled paper effects the environment and natural resources. We suggest that you read this information with your students before making the recycled paper. Students can use information that strikes them as particularly important to make posters or displays to educate the school community. This can be a homework assignment while the class is working on papermaking.

The following facts are excerpted from the *AVR Teachers' Resource Guide for Solid Waste and Recycling Education* (Pedicord, 1984):

If we recycled half of the paper used in the world today we would meet almost three-quarters of the demand for new paper and as a result might arguably free ten million acres of forest from harvest destined for the paper industry. As it stands, we waste about three-quarters of our used paper resource.

According to one estimate (which likely does not assess materials transportation and processing effects), when paper is made from waste paper instead of trees, 60% less water and 70% less energy are used, and 50% less pollutants are added to the environment.

Recycling also helps extend the life of our landfills. Nearly a third of our waste stream by weight and over half by volume is made up of paper. Recycling one ton of this waste saves an average of three cubic yards of landfill space.

These additional facts are from *Oscar Options - Book II* (Bell and Schwartz, 1987):

Paper constitutes at least 50% of the nation's municipal waste by volume.

In an average high school, 85% of the wastes generated are of paper or paper products.

The publication of the Sunday *New York Times* consumes 10,000 trees.

Recycling a stack of newspapers three feet tall saves one tree.

Recycling one ton of newspapers saves about 17 trees.

Recycling one ton of newspapers conserves two to three cubic yards of landfill space.

Further information about the recycling process for paper is from the *AVR Teachers' Resource Guide for Solid Waste and Recycling Education* (cited above):

Wood that is unsuitable for use as lumber and lumber mill wastes are used to manufacture paper. The equivalent of 17 trees are used to produce one ton of paper. Most of this fiber comes from fir and pine forests of the Northwest and pine forests of the South and Southeast. Hardwood forests of the North Central U.S. and New England are also used. 3,688 pounds of wood, 28,000,000,000 BTU's of energy, 216 pounds of lime, 76 pounds of soda ash, 360 pounds of salt cake, and 24,000 gallons of water are used to make just one ton of paper.

To save transportation costs, paper mills are usually located near the forests where the wood is harvested. The trees are debarked, chipped, mixed with chemicals and processed in a large steam-heated pressure cooker called a digester. This helps to break the wood down into cellulose fibers. The fibers are then rinsed with water to remove chemicals, unwanted wood contaminants and dirt.

The remaining water-wood mixture, called slurry, is fed onto a screen and shaken to intermesh the cellulose fibers. Water is drained through the screen and the remaining sheet of paper passes through a series of rollers where it is pressed and dried. The continuous rolls of paper are produced by the machine at a rate of 30 feet per second and can be 16 feet wide.

About 35% of the world's annual commercial wood harvest is used to produce paper, and this share is expected to grow to 50% by the year 2000. Although wood is a renewable resource, we are presently using more of our forest resources than we are replacing. U.S. Government subsidy of lumbering activity keeps the cost of wood products artificially low, encouraging their continued depletion.

In addition to depleting natural resources, manufacturing paper pollutes the environment. According to one estimate, producing one ton of paper creates 84 pounds of air pollutants, 36 pounds of water pollutants, and 176 pounds of solid waste. Disposing of used paper pollutes again, as litter, air pollution from burning, or groundwater contamination from leachate.

Paper is used for three main purposes: 1) as a lightweight, durable surface for writing and printing, 2) to wrap and package food and other items, and 3) for sanitary products such as paper towels and tissue. In 1984, the United States consumed 78 million tons of paper and paperboard products. Of this amount, 43.9 million tons were thrown away, using up more landfill space than any other human-made trash.

Activity Procedure - Part A

1. Several days before the day of production, ask students to brainstorm some of the ways they have heard about to deal with the trash problem. Students will probably mention recycling. In what ways can people recycle? Have students develop a definition for recycling.

2. Explain to students that they will be opening a paper recycling plant. Their job is to make recycled paper and then try to sell their product.

3. To prepare for making paper, ask students to begin saving paper (except newspaper which we found did not work as well) by collecting it in the school. It may take a few days to collect all the paper needed. They may use colored paper if they want their finished paper to be colored.

4. The classroom should be set up so that you have two "production lines" for making recycled paper. Each production line consists of three work stations: one station for shredding paper, the second for blending, and the third for paper-making. Students should be rotated from station to station on their production line so that each person works at every station.

5. Shredding Station: Once enough paper has been collected, have students shred pieces of old paper into a dishpan 1/4 filled with warm water. Some students will continue

shredding paper into the dishpan at the Shredding Station. If possible, soak the paper overnight so it will be easier to blend.

6. Blending Station:

A. Put small quantities of the shredded paper and water (about 4 cups) into a blender. (Too much paper will burn out the motor of the blender.)
B. Blend to a watery, creamy consistency; this mixture is called slurry.
C. Pour the slurry back into the dishpans and continue blending shredded paper until the slurry fills the surface of the dishpan. Add water to the slurry so the dishpan is 1/3 full.

7. Papermaking Station:

A. Place a screen into the water and pull it up horizontally, carefully trying to get the slurry evenly distributed over the screen. If the slurry is too thin, make more slurry of a thicker consistency and mix. If the slurry is too thick add more water.
B. Let water drain from the slurry, then put a paper towel over the pulp. Put newspaper on top and flip over the screen.
C. Use sponges on the screen to soak up excess water. Do not rub sponges against the screen as this will ruin the paper.
D. Carefully remove the screen. If the screen does not come off, continue using sponges to remove water.
E. Use the rolling pin to flatten the paper and make it more even.
F. At this time if you want to add a leaf or other materials to the paper, place the object on the paper and roll the rolling pin over it. The material should stay.
G. Leave the paper on the newspaper to dry. Once it is dry, it will peel from the paper towel.

8. Finished paper can be decorated and made into a greeting card or left plain.

9. Students must now try to sell their product before the next class. Also give them Handout 19, "Recycled Paper - What Is the Story?" to read for homework.

Activity Procedure - Part B

1. Discuss with students:

* What problems did you encounter when trying to sell your paper?
* Was anyone successful in selling the paper?
* If this company was going to stay in the business what would you have to do?
* Does recycling alone help solve the trash problem? Why or why not?

2. In small groups have students make a poster that explains the differences between recycled paper and virgin paper and answer the questions on page 2 of Handout 19.

3. Have each group pick a representative to present their answers to the class.

Suggested Follow-up Activities

Try making recycled paper out of students' newly made recycled paper. Is there a limit to the number of times paper can be recycled? (When paper gets recycled, the wood fibers that hold the paper together get shorter and shorter; eventually they cannot hold the paper together.)

Have students report on recycling efforts in their towns or cities. Is there a recycling program (or a discussion about one)? What items are accepted? What percentage of the local residents participates? Is the percentage increasing? Is participation mandatory? Has the amount of general trash picked up in the area decreased since recycling started? Has the town saved any money? Where are the items given to the recycling program taken, and what is done to them? This activity will give students an insight into how municipal decisions effect residents' everyday lives, a firsthand look at the social studies topic of social change. It will also help students realize that items given to recycling programs don't just go "away" any more than any other trash does.

Have students investigate where recycled products can be purchased in their area. They will likely be able to find several printers who carry recycled paper and even supermarkets that carry various products made from recycled paper. Other items will probably be harder to find, as many recycled resources are used in industry. Have students also look for stores or other places that buy and use recycled goods. For example, are the bags (paper or plastic) at the supermarket made from recycled resources?

Have students start a paper recycling bin for the classroom or the school. Refer to the last activity in this book, "Taking Action," for further ideas for project development and discussion.

RECYCLED PAPER - WHAT IS THE STORY?

Most people assume that paper has always been made from trees. But only since the 1850's have trees been cut and pulped to make paper. For most of its history, paper has been a recycled product.

The first paper, invented in China in the year 105, was made from rags and old fishing nets, and from grasses and plants. The Arabs were probably the first to make paper from used linens. Even North America's first paper mill, built in 1690 in Philadelphia, recycled rags into paper.

Wood fiber from trees was not used to make paper until the 1800's. There was a great demand for paper in Europe then and not enough rags and discarded linens to meet the demand for paper. Inventors started to experiment with other fibers, ranging from swamp grass to marsh mallow to asbestos. Finally, tests using wood pulp produced a usable paper, and at the end of the 19th century the mills switched over to wood pulping techniques.

Today's Recycled Paper

Today we make recycled paper from wastepaper left over in the mills, and from envelope makers, print makers, print shops, homes, and businesses. Paper that has ink on it must be "deinked" to separate the ink from the paper fibers. Except for the work it takes to prepare the fiber, recycled paper and new paper are made the same way.

Why Should We Use Recycled Paper?

Because we can use recycled paper for almost all the same purposes as other paper, we can cut down on the waste, pollution, and energy used to make new paper. We also can help preserve some of our forest lands. Compared to making paper from trees, making a ton of paper from recycled waste paper will save 17 trees, 7000 gallons of water, 60 pounds of air polluting fumes, the energy needed to power an average home for six months, three cubic yards of landfill space, and the money we spend for waste disposal costs.

Recycling paper helps cut down on the problem of solid waste. Research shows that paper fibers can be used up to a dozen times. If we bury the paper in a landfill or burn it in an incinerator, we would lose the chance to use it again and again. We also would use up valuable landfill space.

It is not enough, however, to just collect paper that we have used and send it to be recycled. Recycling programs will just be collecting programs if there aren't many buyers for the paper that is made from recycling.

Doesn't Making Recycled Paper Also Cause Pollution?

We use far fewer toxic chemicals and far less bleach to

make recycled paper than new paper. When a tree is made into pulp, we must remove the "lignin." This is a part of the wood that can interfere with producing high quality paper. Removing the lignin requires toxic chemicals and chlorine. Another toxic chemical is given off when the chlorine interacts with the lignin, adding to air pollution. Also, to make paper white we use bleach. When paper is made from old paper rather than from trees, less bleach is needed since the paper has already been whitened. There is less danger to the environment.

The ink that is removed from paper when it is recycled is usually non-toxic. When it is mixed with the other leftovers (called "sludge") from the recycling process, it can be used by farmers to improve the quality of some types of soil.

Don't All Papers Have Recycled Content?

"Millbroke" is the term used for the scraps of paper that are produced when trees are pulped into paper. These scraps are often collected and used in the next batch of paper making. This helps cut down on the number of trees needed to make paper.

Using "millbroke" is important, but it is different than recycling paper that has already been used in a home, school, or office. Wastepaper from homes, offices, and other places is called "post-consumer waste." This is the paper that would be taking up space and energy in a landfill or waste-to-energy plant. When most people think of paper that has been recycled, they think of old envelopes, letters, and computer

paper being turned into new paper. However, there are no standard definitions for "recycled paper." Some paper that is made from only trees and mill broke is labeled as "recycled," and this does not make a big impact on our solid waste problems. We need to recycle more paper that has already been used by consumers.

Is Recycled Paper As Good As Non-Recycled Paper?

It used to be that recycled paper was not as high quality as non-recycled paper. But today you can get recycled paper in every grade imaginable, including copy paper, computer paper, coated paper, and beautiful text and cover paper. It comes in bright whites and in a wide range of colors, in sheets and in rolls for printer's presses. Strong and versatile, recycled paper meets the same standards as non-recycled paper.

Paper needs a mix of fibers to meet printer expectations. There are some 100% recycled papers that work well on printing presses. Most recycled sheets are a blend of recycled and non-recycled papers. This combination gives you the advantage of both types of fibers. The technical quality doesn't have to suffer.

Not All Papers Are Recyclable

When paper is called "recycled," it means that it has been made from recycled materials. When it is called "recyclable," it means that the piece of paper can be used again to make new paper. Some products use paper combined with other materials -- like envelopes with plastic windows and juice boxes with aluminum foil -- and this makes the paper harder to recycle because the plastic and aluminum must be separated from the paper.

Coated papers, like those used in glossy magazines, are recyclable. Unfortunately, they are not in high demand by the mills that recycle paper. The reason is economic. Coated paper is heavier than regular paper, and a paper mill pays for the paper it gets by its weight. A paper mill gets a lot more sheets of paper -- and the fibers it can use for recycled paper -- from a ton of regular paper than from a ton of paper with heavy coating. Currently, there is a lot of used paper available for recycling, so mills choose uncoated paper. When there are more recycling mills built, they'll need all the used paper they can get, so we may be able to recycle more coated paper.

Why Does Recycled Paper Cost More?

If all paper mills were the same size, actually it would be less expensive to make and sell recycled paper. But bigger mills have bigger machines that can make paper more quickly and more cheaply, and today the biggest mills still use trees; consequently, virgin paper is cheaper.

Only recently have we been collecting enough used paper to consider building large mills for recycling. Also, we have only recently learned ways to remove the ink from some types of papers, like photocopy paper. While recycled paper is often still more expensive than non-recycled paper, the difference in price is getting smaller, and someday recycled paper will likely cost less than non-recycled paper.

What Does the Future Hold For Recycling Paper

Right now we are collecting more newspapers for recycling than are being used. Many companies and individuals are not buying and using recycled paper. Until recycled paper is bought, there won't be a lot of progress. If you or a company is not buying recycled products, then you (or they) are not recycling.

State governments have started to use recycled papers. They have made a small impact on the papermaking industry. The federal government uses only 2% of all the paper bought in our nation. What the government uses just can't make a huge difference and change the ways paper is made. If we are going to make progress in using more recycled papers, it will have to come from major corporations and printers, who use the most paper.

Throughout history, the paper-making industry has seen many changes. Some of the changes resulted from economic reasons, some from the need to use new materials , some from laws and regulations. Today our environmental needs may force yet another change. A product as ordinary as paper may provide a powerful step toward living in a more environmentally sound way.

Name _____

REFLECTING ON RECYCLING

1. Why might recyled paper cost more?

2. Why did you have problems selling your paper or why were you able to sell your paper?

3. How could you make your paper more marketable?

4. If you were going to invest money in a factory, would you invest in your student operation? Why or why not?

5. According to the reading, what will make recycled papermaking plants be more successful?

6. Based on the reading, how would you change your papermaking plant to be more successful?

Activity 13

SOURCE REDUCTION: BROWN BAG LUNCH

Objective

Using their own lunches, students will learn about source reduction and apply the principle to their own behavior.

Note

For some students, bringing lunch from home is problematic, so you will have to be sensitive to the individual circumstances of your students. In communities where most students eat school lunches, this activity may not be appropriate.

Suggested Previous Activities

2. Changing Times, Changing Waste
10. Do You Know Where Your Trash Is?
11. What Should Be Done with All the Trash?
12. Recycling: Making Paper

Time

5 minutes to introduce the activity
20 minutes the following day to examine lunch trash
30 minutes the next day to compare trash and discuss homework assignment (bag lunches on 2 consecutive days)

Vocabulary

compost	source reduction
recycling	trash
reusing	

Materials

paper and pencils
each student's trash from lunch on 2 consecutive days
1 copy of Handout 20, "Brown Bag Lunch," for each student
1 copy of Handout 21, "Reducing, Reusing, and Recycling," for each student

Supplementary Information for Teachers

The Environmental Protection Agency's recommendation for all communities is that they use an "integrated waste management" approach to dealing with their garbage. Integrated waste management means using a combination of practices that will result in the least impact on human health and the environment. Most management systems will use some or all of these components: source reduction, recycling, waste-to-energy (resource recovery) facilities, landfills.

Source reduction involves decreasing the amount of trash generated. This can be done by selective buying habits and reusing products and materials rather than throwing them away. For example, a consumer can choose to buy fruit either packaged or unpackaged. If the consumer buys it unpackaged, he or she is practicing source reduction.

Activity Procedure

1. Explain to students that on two consecutive days they should bring lunches from home (not purchase school lunch) and save the trash from each lunch. Remind them of what is included in the definition of trash.

2. After the first day that the students bring their lunches, have each student list all the items in the trash from her or his lunch on Handout 20. Remind students to include the bag itself if it is not reused.

3. In small groups of three or four have students:

    * find the most common item thrown away
    * agree on a definition of source reduction
    * brainstorm ways they can reduce the amount of garbage generated from one lunch (what disposable items might be replaced with reusable items, e.g., juice box with a thermos, brown bag with a lunch box or cloth bag, etc.)

* Have one student in the group record the group's answers to share with the rest of the class.

4. Have the spokesperson from each group share the answers. The class should come up with one definition of source reduction. You might also list ways to reduce trash from lunch on the board.

5. Collect all the trash from all the students' lunches and put it in a box or bag to be compared with the next day's lunch trash.

6. Give students the homework task of coming up with lunches for the next day that will generate the least amount of trash.

7. Have students bring trash from the next day's lunch to class.

8. Have students fill out remaining questions on Handout 20.

9. In the same small groups as the preceding day, have students discuss answers to handout questions, and ask the group to choose one spokesperson who will report the group's findings to the class during the class discussion.

10. Collect the day's lunch trash from all students in a container like the one used for the preceding day's trash. Note the difference in weight and volume of the two days' trash.

11. Conduct a discussion with the entire class. Have the spokesperson from each group report the group findings.

   * Was the class able to reduce the amount of trash? By how much?
   * Did the difference surprise you?
   * Is source reduction a good way to reduce trash? Why or why not?
   * What problems does source reduction pose?
   * How might any problems be resolved?
   * What was the most common way we reduced our trash?

12. Continue the class discussion by having students compare their experience of making recycled paper with their experience of reducing lunch waste. Also, give students Handout 21, "Reducing, Reusing, and Recycling."

The bag lunch activity is a chance for students to understand the term *source reduction*, that is, how amounts of trash can be reduced. Through papermaking, students learned about *recycling*; another important term is *reusing*. Though the terms *reuse* and *recycle* are used broadly and sometimes interchangeably, it is important that students consider the difference between using an item repeatedly in its original form (like a cloth shopping bag) and using the ingredients of an item after changing its form (like making recycled paper). The latter requires energy and produces some waste, though less than using something brand new. Both are ways to reduce the amount of trash sent to landfills and incinerators.

   * Besides making paper, what are other examples of recycling? (Students might think of companies or towns that take glass, plastic, or aluminum and clean and melt or shred them to form new items. Other examples might be shredding paper to use it as packing material, making a patchwork quilt from old clothes, etc.)
   * Those students who put their lunch in the same bag both days were reusing the bag. What else can be reused? (Students might list toys or clothes given to younger siblings, wrapping presents in newspaper, cleaning a jar to use as a pencil holder, etc.)
   * List some of the ways that PC Connection recycled, reduced, or reused resources.
   * How are all of these practices ways of reducing waste?

Suggested Follow-up Activity

Have students bring an object from home that would ordinarily be thrown out and find another use for it. (Examples: plastic containers to grow plants or start seeds, cans or plastic containers for letter holders.)

Name _____

BROWN BAG LUNCH

1. Categorize and list below all the items in your lunch trash from the first day.

2. Categorize and list below all the items from your lunch trash from the second day.

3. What methods did you use to reduce the amount of trash you produced?

4. What trash was easy to reduce? What was difficult to reduce?

5. What changes in habit were required to reduce the amount of trash in your lunch?

6. Is source reduction such as you did in this experiment something that you could continue? Explain.

7. Write a short statement summarizing your findings about source reduction from this experiment.

REDUCING, REUSING, AND RECYCLING*

The statements below are excerpts from a brochure from the company PC Connection.

Why we're putting less in every box.

These days a great way to give more is to give less. So we've developed a way to ship safely using less packaging. And we've eliminated styrofoam peanuts.

Early on we switched to a non-static foam peanut without hard CFC's. But as the environmental case against foam products grew, we decided to switch to recyclable newsprint.

Just as important, we're reducing our *volume* of packing materials by using a greater variety of box sizes. Now every order fits more tightly and needs little (if any) extra packaging.

We're finding ways to send less to the dump, too.

...We had been doing some recycling already. But truckloads of garbage were still going to the dump. So we started looking for markets for all those thousands of cubic feet of corrugated (cardboard), hundreds of pounds of steel banding, and huge stacks of miscellaneous styrofoam and plastic. We also began sorting office paper for efficient recycling and making better use of electronic mail (and other forms of source reduction). Overall, in just four months we've reduced waste by 75%.

We even have a compost pile behind the utility shed so food scraps from our lunchroom can be transformed into delicious dirt for the local flora....

* Reprinted by permission of *PC Connection*

Activity 14

PAPER OR PLASTIC?

Objective

In this activity students will explore the concept of source reduction. By examining their own and other's' choices of shopping bags students will experience the problems related to decision-making about source reduction. They will also learn that environmental problems are created by using either paper or plastic.

Suggested Previous Activities:

1. How Much Trash Do You Throw Away in a Week?
2. Changing Times, Changing Waste
10. Do You Know Where Your Trash Is?
11. What Should Be Done with All the Trash?

Time

15 minutes to introduce the assignment
1 hour homework (survey)
30 minutes homework (reading)
15 minutes for follow-up discussion

Vocabulary

aerobic
bacteria
carcinogens
degradable
hazardous waste
herbicides
lignin
microrganism
monocultures

nonrenewable
nutrients
recyclable
recycling
renewable
reusing
source reduction
toxic

Materials

1 paper shopping bag
1 plastic shopping bag
1 cloth shopping bag
5 - 6 copies of Handout 22, "Supermarket Survey on Bags" for each student
1 copy of Handout 23, "Bag It" for each student

Activity Procedure

1. Bring in samples of various shopping bags and ask students which kind of bag they most commonly get from a store.

2. Have students answer the survey questions on Handout 22 in small groups.

3. Explain the homework assignment: Students should go to a supermarket or other retail store and ask the survey questions to four or five different shoppers. If they cannot get to a store, they can ask other teachers, neighbors, or parents. You may want to have the students practice interviewing techniques in class.

4. After students have done the survey, they should also read Handout 23, "Bag It," for homework.

5. Hold a class discussion to help students draw conclusions from their surveys and the article. In small groups or as a class, ask students these questions:

* What responses did you receive the most?
* Were you surprised by the responses you got? Why or why not?
* What will you use when you go shopping now? Why?
* How do you think people form their opinions about paper versus plastic?
* How do the thoughts from consumers about the advantages of paper versus plastic bags compare with what you learned from the "Bag It" article?
* Is the public well-informed? Why or why not?

SUPERMARKET SURVEY ON BAGS

1. Do you prefer paper or plastic bags when you shop? Why?

2. What do you think are the environmental effects of using plastic bags?

3. What do you think are the environmental effects of using paper bags?

4. What do you think is the most environmentally sound way to transport your groceries?

5. What would you think if stores stopped giving out bags?

6. Do you ever bring your own bags to the store?

ENVIRONMENT/
PHYSICAL SCIENCE

Standing at the grocery store checkout counter, you are faced with a choice: paper or plastic to pack your purchases. Which bag is more environmentally friendly? You might be surprised.

by Karen McNulty

hances are you're an environmentally conscious shopper, so you grab for the paper bag, right? After all, *everyone* knows that plastics are not *degradable* (able to break down in nature) while paper bags are. That alone, you might say, is a good enough reason to choose paper over plastic. Paper would at least cut down on litter on the street.

But you'd better not act so fast. When you get right down to it, figuring out which type of bag is less harmful to the environment is not such an easy task.

For one thing, you've got to keep in mind that *both* types of bags pose many environmental dangers—even before they become bags. That is, the *manufacture* of both paper and plastic creates many pollutants. To make a truly informed decision, you have to follow both kinds of bags from beginning to end, and consider all of the potential problems.

After reading through the chart and the accompanying articles on pages 12-16, you'll be better armed with facts to help you decide just how to pack your groceries.

BAG IT

THE GROCERY SACK DILEMMA

HOW PAPER BAGS ARE MADE

1. Paper is made from trees. Well, actually, wood fibers that were once a part of living cells in tree trunks. The longer the fibers, the stronger the paper. Since brown paper bags have to be strong to hold groceries, they are usually made from long-fibered softwood trees, such as pines. Chopping down these trees disrupts wildlife habitats.

2. Trees are cut into logs and carried to a paper mill by truck or on a flowing stream. The truck uses energy and creates pollutants, but the natural route has problems too. The logs can overload the stream with nutrients and chemicals. Algae and other microorganisms thrive on the nutrients and multiply, using up the stream's oxygen supply. Some of the chemical pollutants also deplete the oxygen supply. Fish and other life-forms that use oxygen may die as a result.

3. At the mill, a revolving *barking drum* removes bark from the logs.

4. Then a *chipper* cuts the wood into small chips.

5. The wood chips are converted into a soft *pulp* (like mashed potatoes) by cooking them at a very high temperature with *sodium hydroxide* (lye) and *sodium sulfide*. These chemicals digest the *lignin*—the natural glue that holds the wood fibers together in the tree—so only the fibers remain.

6. The pulp is now a watery mix of chemicals and fibers. It is *refined* to separate the individual fibers. Then it is washed to remove the chemicals. The waste water is usually treated to remove some of these chemicals before it runs into a nearby stream. But many chemicals remain, which further deplete the stream's oxygen.

HOW PLASTIC BAGS ARE MADE

1. Plastics begin as oil or natural gas. Both oil and gas are taken from below the earth's surface, where they occur naturally but in limited quantity. Drilling for these resources may disrupt wildlife habitats. And, as today's reserves are used up, we will need to explore new areas, disrupting more wildlife.

2. Transporting oil and natural gas from the drilling sites requires energy. It also presents the possibility of environmental damage from accidental spills, like that from the *Exxon Valdez*.

3. At the refinery, oil and natural gas are refined into heating and transportation fuels. *Ethane gas*, a waste product from this process, is used to make plastics and other chemicals. (Otherwise it is burned off.)

4. The ethane gas is heated so that a chemical reaction takes place to convert it to *ethylene*, a chemical that is the building block of many plastics. This process creates many hazardous wastes, but most are burned as fuel to keep the process going.

5. When combined with other chemicals, the ethylene building blocks join together to form *polyethylene*—the plastic from which grocery bags are made. The polyethylene may be mixed with addition-

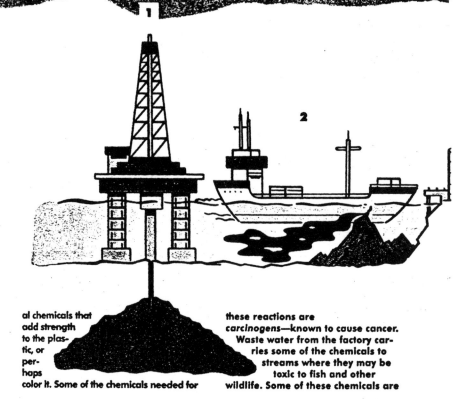

al chemicals that add strength to the plastic, or perhaps color it. Some of the chemicals needed for these reactions are *carcinogens*—known to cause cancer. Waste water from the factory carries some of the chemicals to streams where they may be toxic to fish and other wildlife. Some of these chemicals are

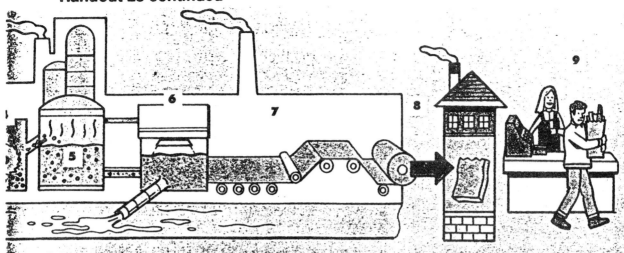

7. The pulp is then spread across a moving screen. More waste water drains off, leaving a mat of fibers on the screen. A series of rollers squeezes and dries the sheet of paper before it is stored on huge rollers.

It takes a lot of electricity to run the machines needed for papermaking. This electricity is generated by burning fossil fuels (usually coal) and wood. This

burning releases many pollutants into the air, including *sulfur dioxide* and *nitrogen oxides* (the chief components of acid rain), as well as *carbon monoxide*, and *dust*. Because sulfur compounds are used in the pulping process, *hydrogen sulfide* gas makes the air smell like rotten eggs.

8. At the bag factory, machines use

more electricity to cut, fold, and glue the paper to form bags. Some of the glues and the inks used to print on the bags may contain toxic substances.

9. At the grocery store, customers pack their groceries in these brown paper bags and carry them home.

Brad Hamann

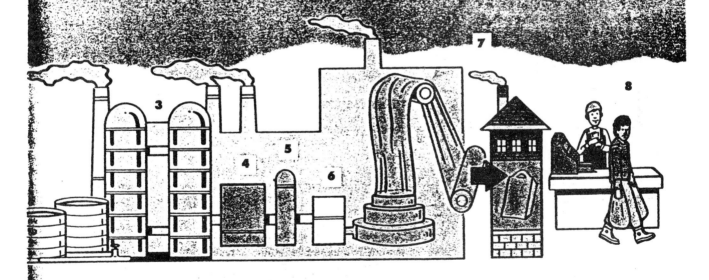

also released into the air.

6. Polyethylene is heated to form a hollow tube, and blown up with air like a balloon. The double sheet of polyethylene is then stored on large rollers.

All of these processes require energy obtained from burning fossil fuels such

as oil. Pollutants, including *sulfur dioxide, nitrogen oxides, carbon monoxide*, and *dust* are released into the air.

7. At the bag factory, electricity is used to run machines that cut, shape, and seal the polyethylene sheets to form

bags. Inks used to print on the bags may contain toxic chemicals, such as lead and cadmium.

8. At the grocery store, customers pack their groceries in these polyethylene bags and carry them home.

Starting to feel confused over which pack is a better environmental pick? Better get some more details.

Brad Hamann

Both paper bags and plastic bags may end up in a dump. Plastic takes up less space, but paper degrades faster.

Many people feel the case for choosing paper over plastic is clear. After all, plastics come from nonrenewable resources—oil and natural gas. And plastics create pollution, both during their manufacture and when their useful life expires.

Good ol' brown paper bags, on the other hand, have been around for ages. They're durable yet degradable—even recyclable. And they come from a resource that can be replenished—trees. What could be wrong with a product that's so obviously 100 percent natural?

Plenty, according to a West German study designed to compare the overall environmental impact of the two kinds of bags. In fact, *some environmentalists now say that brown paper bags cause more harm to the environment than polyethylene plastic bags.* Here are some points to consider.

THE RAW MATERIALS

Drilling for oil and gas does damage the environment in some cases. Last year's spill from the *Exxon Valdez* killed thousands of sea birds and otters in Alaska's Prince William Sound, for example. And environmentalists fear that continuing to make plastics will speed up the depletion of our limited reserves of oil and gas. When the old reserves dry up, the petroleum industry may begin drilling in more sensitive areas, such as the Arctic National Wildlife Reserve on Alaska's northern coastal plain (see *SW*, Dec. 18, 1987, p. 6).

But petroleum industry representatives say that very little oil and gas actually go into making plastics. Eliminating the use of these materials for making plastics would not increase the life of the reserves, they say.

In fact, says Sidney Rankin, a chemical engineer at Rutgers University, the plastics that do come from oil are made from *ethane gas.* "That's what's left over after you make fuel during the oil refining process," says Rankin. "It used to be burned off as waste."

Some environmentalists admit that making plastics out of ethane is actually a good use of what was once a waste product. In some ways they believe it is better than cutting down trees.

"Cutting down trees to make paper destroys wildlife habitats," says Jan Beyea, senior policy analyst for the National Audubon Society. Trees are renewable, and the paper companies do replant their forests, admits Beyea. "But they don't replant the kind of forests that were there before," he says. They plant *monocultures*—forests that grow only one kind of tree. Without diversity in the forests, says Beyea, "some creatures that used to live there can't."

Beyea also suggests that constantly cutting and replanting trees, along with using herbicides to kill weeds, will eventually deplete the forest soil of nutrients. Then even the trees won't be able to survive.

The paper industry can solve these problems and develop ways to make brown paper bags out of recycled paper, says Beyea. But according to David Stuck of the American Paper Institute, "The bags wouldn't be strong enough if we used recycled paper using current technology. The more you recycle paper, the shorter the fibers become," he says. "And you need long fibers to carry the weight of the groceries."

MANUFACTURE

It takes a lot of energy and a lot of chemicals to make both paper and plastic. And both processes generate a tremendous amount of waste (see diagram, pp. 12-13).

But, according to the West German study that compared the production processes for the two materials, it takes 50 percent *more* energy to make 50,000 brown paper bags than it takes to make the same number of polyethylene bags. That means paper mills have to burn more fossil fuels (mostly coal) than plastics factories do to generate electricity. In addition, the study found that making paper bags creates more air and water pollution than making plastic bags.

"Paper mills are notorious polluters," says Dorene Bolze, another environmental policy analyst at the National Audubon Society. According to the German study, they release twice as much sulfur dioxide as plastics factories do, and almost 50 percent more nitrogen oxides. Both of these chemicals contribute to the formation of acid rain. The paper mills also release more carbon monoxide and dust into the atmosphere than plastics factories do. And unlike plastics factories, paper mills emit sulfur

compounds, such as hydrogen sulfide gas, which make the air smell like rotten eggs!

In addition, the waste water from paper mills is generally much more polluted with oxygen-depleting chemicals than waste water from plastics factories. These chemicals "choke the life out of the streams," says Bolze, because fish need oxygen to live.

Plastics factories do emit more hydrocarbons (compounds made of hydrogen and carbon, such as *benzene*) into the air and waste water. These chemicals may cause cancer or poison wildlife, says the Audubon Society's Beyea.

DISPOSAL

Since the paper bag is *degradable*—bacteria can digest its fibers in the presence of water and air—it *must* score environmental points here, right? Well, yes—and no. You see, it depends on where the bag ends up.

If packed into a sanitary landfill, as about 90 percent of our garbage is, then it really doesn't make a difference if the bag is degradable or not. Landfills are packed so tightly that the *aerobic* (oxygen using) bacteria that normally degrade garbage can't survive. Scientists have dug up plastic containers, newspapers, and even vegetables that have been buried in landfills for 10 or 20 years. None of them were degraded. So if nothing degrades, plastic bags can't hurt. Some say they may even help the situation. Plastic bags are less bulky than paper ones, so they

Bill Stanton

In a landfill (*below, left*) it may not matter if your trash is degradable. But as litter, paper fades away while plastic stays and stays.

may save some landfill space.

But if you're talking about garbage that misses the mark and ends up as litter on the street or at sea, then paper is definitely the better choice, says Audubon's Beyea. Plastic bags do not degrade, even in the open air. On land, that makes for ugly scenery. In the ocean, plastics can kill. Sea turtles, for example, think floating plastic bags look like jellyfish—a good meal. Unfortunately, the turtles can't digest their plastic dinners, and they die. Hundreds of thousands of sea birds and marine mammals die each year when they eat or get tangled up in plastic waste. A paper bag, in contrast, will degrade within a week once it gets wet.

So, if you *must* make a choice between the two types of bags, think about where you are. "If you're using the bag near the coast, where it may enter the water, always choose the paper bag," says Beyea. But if you're inland and careful about what happens to your trash, the plastic sack is a better pick.

Barbara Hansen

Bringing your own bag saves raw materials and energy, and helps cut back on pollution and waste.

So what if you still can't decide. Don't. Just B.Y.O.B.—bring your own bag!

By now you've realized that picking between paper and plastic bags can get to be pretty tricky. And maybe you've come to the realization that what the bags are made of isn't the problem after all. It's simply the fact that they're *both* made to be thrown away. So, if you really want to help the environment, and avoid the plastics vs. paper debate in the process, learn how to *recycle*.

That doesn't mean you have to set up a messy recycling factory in your kitchen. All you have to do is use your grocery bags—whether they are paper or plastic—again and again. Each time you reuse one bag, you save the raw materials, energy, pollution, and contribution to trash that would have resulted from throwing it away.

Here are some suggested uses for old paper and plastic bags:

PAPER BAGS
- Bookcovers
- Masks
- Packing newspapers for recycling

PLASTIC BAGS
- Beach bags for wet bathing suits and towels
- Trash basket liners
- Packing a picnic lunch

The most direct way to recycle both kinds of bags, however, is to bring them back to the grocery store so you can tote home another load of groceries. That way you won't need to take another bag—or make the choice between paper and plastic.

Better yet, why not bag both the paper and plastic? Do like many Europeans do. Get yourself a mesh bag that is built tough to carry groceries over and over again.

The West German environmental scientists checked out reuseable bags that could be used up to 100 times each, and compared them to throwaway plastic bags. Since each reuseable bag can do the job of 100 throwaway bags, only one has to be made for every 100 plastic bags. The West German scientists found that making that one reuseable bag uses one-tenth the amount of the energy it takes to make the 100 plastic bags. And making the reusable bag creates only one-tenth as much waste.

So if you really want to do something good for the environment, bring your own bag.

And try to think of other reuseable items you can substitute for common throwaways. Otherwise, sooner or later, we may be buried in our trash.

REACTION TIME:

1. Why is it important to consider the raw materials, manufacturing process, *and* waste-disposal options when deciding which type of bag is more damaging to the environment?

2. Could you use the comparisons made in this article to decide if food packages (such as fast-food hamburger packages) should be made out of polystyrene (a different kind of plastic) or paperboard (a different kind of paper)? Why or why not?

3. Would banning all plastics be good for the environment? Explain. ■

Activity 15

PACKAGING

Objective

In the process of evaluating and creating product packaging, students will become aware of some of the reasons for packaging, the effect it has on the volume of the waste stream, and some ways to reduce it.

Suggested Previous Activities

2. Changing Times, Changing Waste
10. Do You Know Where Your Trash Is?
13. Source Reduction: Brown Bag Lunch
14. Paper or Plastic?

Time

45 minutes to introduce the topic, and look at Handout 24, "The Good, the Bad, and the Ugly"
homework (reading Handout 25)
45 minutes to discuss Handout 25, read and discuss Handout 26)
homework to find and design packaging (might take more than one night or class session)
45 minutes to display and discuss designs
45 minutes class time or homework to start letters to manufacturers and /or legislators (completing them might take more than one class session or night)

Vocabulary

aseptic
biodegradable
non-renewable resources
photodegradable
recycled

recyclable
renewable resources
reusable
source reduction

Materials

examples of packaging that vary in degree of efficiency
1 copy of Handout 24, "The Good, the Bad, and the Ugly", for each student
materials at home for students to use to design packaging such as cardboard, styrofoam, string, paper, bags
1 copy of Handout 25, "What's Wrong with a Pretty Package?," for each student
1 copy of Handout 26, "Maine and the State of the Juice Box," for each student
paper and pencils

Activity Procedure

1. Explain to students that approximately one-third of solid waste is packaging, and thus deserves attention. Bring to class several items that will serve as examples of good, mediocre, and bad packaging. Some examples might be: an ice cream cone, an egg carton, a compact disc in its wrappings, an aspirin bottle in its layers, and gift boxes of candies. Have students look at these items, and ask questions to help them learn about packaging.

> * Why do you think manufacturers use packaging? (advertisement; safety, as with medication; portion control, etc.)
> * Look at these items and how they are packaged. Do you think they are effectively packaged? If yes, in what way? If no, why not?
> * What kinds of packaging are important? What kinds are extraneous and wasteful?

2. Have the students list criteria for good packaging and bad packaging. Write these ideas on the board.

3. Distribute Handout 24, "The Good, the Bad, and the Ugly." With the whole class, look at the examples included in this article on packaging. Ask students to compare the article's criteria with the list they generated.

4. Have students read Handout 25, "What's Wrong with a Pretty Package?" for homework.

5. In class the next day discuss what they learned about good and bad packaging. Then have students read Handout 26, "Maine and the State of the Juice Box." Use the questions below to discuss the article in class.

> * What encouraged the research into recycling of the juice boxes?
> * Why might some people argue that the juice box is recyclable?
> * Why might some people argue that the juice box is not recyclable?
> * Why did legislators in Maine ban the juice box?
> * Who influences legislators? (Encourage students to brainstorm the many ways to pressure government.)
> * Are there other examples of industry being affected by legislation? (In Massachusetts newspaper publishers agreed to use more recycled newsprint when the Massachusetts legislature was considering a bill to make it a requirement; returnable bottle bills have been enacted in several states).

6. Give the following homework assignment: Have students go to the grocery or drug store and find an item that they think is over-packaged. Ask students to write down why they think the item is over-packaged and explain how they would package it differently. Students should make their own alternative packaging and bring it to class. (Students might need more than one night or some class time for this task.)

7. Discuss the students' package-making process using these questions:

> * What environmental concerns did you take into account when you made your decisions about packaging?
> * How was your packaging different from the manufacturer's packaging?

* Did you evaluate packaging according to whether it was made of renewable resources or nonrenewable resources?
* Why do you think your product was originally packaged the way it was? Are there important reasons for using this particular type of packaging?

8. In class and for homework, students can do one of the following assignments. These tasks might take a few class periods.

A. Write to the companies that manufacture the products for which students redesigned the packaging to find out why the companies used the packaging that they did. The students can suggest alternative packaging based on their own research and/or information from the articles. They can explain why they feel that the alternative packaging is important. You may need to discuss with students the tone of their letters, the importance of revision and careful editing, and ways to find addresses from the products' labels.

B. Write to legislators to encourage then to write legislation that will help with the packaging problem. They might want to think about legislation that requires companies to use only recycled materials in their packaging, or that limits the number of packaging materials.

THE GOOD, THE BAD, AND THE UGLY

GOOD: Tom's Toothpaste

Our choice for smart toothpaste packaging is the Tom's of Maine line, the only nationally available toothpaste packaged in a 100 percent aluminum tube in a recycled paperboard box. Most other tubed toothpaste — a disappearing breed — comes in plastic laminate, virtually impossible to recycle.

Tom's packaging makes even more sense in light of the trend toward polypropylene/polyethylene pump dispensers, also virtually unrecyclable. What's more, the high-tech containers can't be manipulated to dispense every ounce of paste, so they have to be bought — and thrown out — more frequently than the tubes.

GOOD: The Castle Crate

Manufactured by California-based Rehrig Pacific, the Castle Crate containers for two-liter polyethylene terephthalate (PET) soda bottles were designed for a dual purpose. Each high-density polyethylene crate holds eight bottles for shipping and serves as a display once soft drinks reach retailers. Crates safely lock into each other during transport and eliminate the need for cardboard shipping boxes and corrugated paperboard filler. The crates have a seven- to 12-year life span so they can be used again and again. Both Coca-Cola and Pepsi have begun to use them.

BAD: Compact Discs

Here packaging manufacturers have gone overboard with the paperboard. Most CDs come in a hard plastic case for permanent storage — no complaints there. But do we really need the six-inch-by-twelve-inch paperboard box — twice the size of the CD — plus an additional layer of shrink wrap? (Some CDs come sandwiched in polyvinyl chloride blister packs, also unnecessary.) Manufacturers favor the extra-long box as a theft preventive. But CDs could easily be stacked in locked glass cases, eliminating the need for the oversized box.

BAD: PVC Bottle

It's getting so that you can hardly find vegetable oils in glass bottles. Many oils and many bottled waters now come packaged in polyvinyl chloride. The Federal Food and Drug Administration is currently preparing an environmental-impact statement on this resin, which contains chlorinated compounds and is believed to contribute to dioxin and furan emissions when the bottles are burned in incinerators. The environmental effects of two plasticizers, DEHA and DEHP, used to give PVC film flexibility are also being examined.

BAD: The Fast-Food Clamshell

The McDonald's Corporation has given its widely criticized polystyrene hamburger container an image rehaul, but the corporation hasn't eliminated its environmental problems. The clamshells are no longer made with CFC-12, a compound linked to destruction of protective ozone in the upper atmosphere. But they are made with a questionable substitute, HCFC-22. Some containers are blown with the hydrogen-based compound pentane. The thickness of the clamshells has been reduced by

12 percent. And last fall the company launched a $16 million experimental recycling program to reclaim the containers and convert them into plastic pellets which could then be turned into trays, toys, and videocassettes. But the vast majority of polystyrene clamshells still end up at the dump.

GOOD: Refillable Bottles

The 16-ounce Grolsch beer bottle sold individually or in pairs is refillable and looks good, too. The import from Holland comes with a wire-attached ceramic plug rather than a tossable metal cap. The handsome dark-green bottle doubles as an air-tight container for oil, vinegar, or salad dressing. Beverages in refillable bottles are even better for the environment than recyclable ones, as sterilizing and refilling are less energy-intensive than reprocessing. Here's hoping the Euro-style bottle sweeps America.

Reprinted by permission of *Garbage Magazine*.

UGLY: Squeezit

For true packaging excess, scan the supermarket aisles for kids' snack foods. You'll find single-serving packs of chips, puddings, and drinks. A newcomer to the kids' food line is Betty Crocker's Squeezit from General Mills, a 6.75-ounce squirtable juice product sold in polyethylene six packs. As if the consume-and-toss containers weren't bad enough, they're held together with a paperboard yoke and plastic wrap.

GOOD: Welch's Dinosaur Jelly Glasses

After a ten-year absence from store shelves, Welch's has re-introduced jelly jars that double as drinking glasses once the jelly's gone. This is smart packaging at its best — all that's tossed is the lid. Product information that would otherwise be printed on a paper label is provided on the cap. (Some Welch's jelly jars from the 1950s are now collectors' items.)

BAD: Aunt Jemima's Pancake Express

Aunt Jemima's Pancake Express from Quaker Oats is one of the worst convenience products we've run across. Packaged in polypropylene, the 4.5 ounces of prepared pancake mix go straight from container to grill when water is added, eliminating the need for a mixing bowl and spoon. The real clincher is that the Express mix makes just six pancakes — or breakfast for two. So unless polypropylene is getting recycled in your community in any significant quantity — unlikely at best — the Express mix container goes straight to the landfill or incinerator after one morning's use.

BAD: Microwavable Potatoes

Now microwavable side dishes are becoming chic. Case in point: Club Chef Incorporated's whole fresh potatoes packaged on a plastic tray and covered in shrink wrap. Once the film is pierced, the whole unit can get zapped. The plastics can be tossed seconds later. Pulling a potato out of a five-pound bag and zapping it on a pyrex plate *can't* take that much longer.

GOOD: Recycled Spic and Span Containers

Procter & Gamble deserves a pat on the back for its Spic and Span pine-scented liquid cleaner in 100-percent recycled polyethylene terephthalate (PET) bottles. Right now the product is available only on the West Coast, but P&G is expanding distribution. Until this breakthrough, PET bottles could only be recycled into lower-grade products like plastic flowerpots and stuffing for ski jackets — not bottles — meaning more virgin petroleum was needed to make new bottles.

GOOD: Tools in the Nude

Remember when tools were hung from their handles on pegboard in hardware stores? Today, tools, nails, screws, and bolts often come encased in bubble-like blister packs which get ripped open and can never be used again. Thumbs up to shops that *still* hang pliers, wrenches, and hammers and keep nails, screws, and bolts in bins.

GOOD: Tide with Bleach

Detergents of varying strengths, blue-ings, bleaches, fabric softeners, spot removers, wool cleaners — they all add up to a lot of garbage. Procter & Gamble's Tide with Bleach cuts down on packaging by combining two products (detergent and bleach) in one box. P&G also uses recycled paperboard for the container. What's inside is also improved: Tide with Bleach uses environmentally safer non-chlorine bleaches. Now if P&G would just get rid of the phosphates, we'd give the product unqualified kudos.

BAD: NyQuil

No one denies the need for tamper-proof packaging, but NyQuil packaged in a plastic bottle, safety-sealed with a rigid cap, and boxed in paperboard is a bit much. Manufacturers could easily dispense with the box and maintain the safety of the product. Also unnecessary is the plastic measuring cup accompanying the cold medicine. A spoon does the job just as well.

— GINIA BELLAFANTE AND JANET MARINELLI

WHAT'S WRONG WITH A PRETTY PACKAGE

The next time you go to the grocery store, look at all the different packages. They have bright colors, pictures, and sometimes games, all designed to make you notice the product and buy it. These packages, while fun to look at, are also causing a problem. According to "garbologists" who study garbage, up to 65% of what we throw out is packaging.

Would you be willing to buy your favorite cereal or candy bar even if it didn't have a fancy box or wrapper? Many American shoppers say they would, even if it cost more. They would want their products sold in what is called "environmentally friendly" packaging. This is packaging that can be made from recycled papers, or the package can be recycled, or the package can be used at home when you finish the product inside. Some environmentally friendly packages are also "biodegradable" -- they break apart and return safely to the earth and air.

Many companies are starting to make environmentally friendly packages. The trouble is that the plans for these packages don't always work out right. Some of the degradable packages, such as photo-degradable plastics, are ending up buried in landfills where they will never get a chance to break apart. They'll just stay buried, taking up room, for years and years.

Some packages say that they can be recycled, but there is no one who can recycle them! Or the recycling plant is a thousand miles away. Just because a package says that it can be recycled is no guarantee that it will be.

Things get even more complicated if you think about what it takes to make a recyclable package in the first place. How much air and water pollution were produced to make the package in the factory? Does the package need to be refrigerated? If so, that uses a lot of energy. Do trucks and trains burn up a lot of fuel driving the packages from place to place? As many experts are beginning to say, environmentally friendly packages are just one solution to our growing trash problem. What we really need to do is use less packaging right from the start.

Making the Best Choices

Using less packaging from the start is known as "source reduction." It means doing things such as bringing your own reusable bag to the supermarket, rather than using new plastic or paper bags every time you shop (see the article "Bag It" in Activity 14). For companies that make products that need packaging, it means doing it in the safest and most environmentally friendly way possible.

To encourage the companies to do this, the governors of nine states got together and came up with a proposal to cut package wastes.

They want laws passed that force companies to use safer inks and dyes. They also want companies to remove certain harmful chemicals in packages that pollute the air when they get burned in incinerators.

The governors are trying to come up with guidelines that will help all of us use and buy the best packaging. From best to worst, here is what they'd like to see:

No Packaging: Some products don't need a package. For instance, why wrap a hammer in a "blister pack" (one of those see-through plastic covers attached to cardboard) when it can just hang on a rack? Many products will do fine unwrapped.

Minimal Packaging: Some products need just a little bit of packaging. Some boxes can be made from thinner cardboard. Concentrated liquids, such as some laundry detergents or frozen juices, can use less packaging. Combining the laundry detergent with bleach in one package is even better.

Reusable Packaging: Companies can ship their products in large containers that can be sent back to the factory. Jelly jars can be kept at home to be used as drinking glasses.

Recyclable Packaging : While it is good to be able to recycle a product, we need places to do it. But, if a product needs a certain type of packaging, recyclable packaging is better than just expecting the package to become trash.

Why Change Takes Time

A hundred years ago, it was very fashionable to go to the market with a fancy box or tin can of your own, fill it up, and use it again and again. Now, people have different habits. Microwave ovens are very popular, and the food we buy for them has a lot of packaging. Medicines come in special packages that children cannot open. Our world is very complex and filled with packaging.

Many people also work for companies that design and sell all those packages. Supermarkets and drugstores often depend on the packages to persuade people to buy their goods. They may not want to change their packaging unless they are sure they can still sell their products. Some designers think they can make interesting packages that are environmentally friendly, and the shoppers will appreciate this. It will take much effort, and perhaps money, to change the habits of businesses and shoppers.

Some Companies Are Trying

The Proctor and Gamble Company (P&G) is working hard to reduce the packaging it uses. It hopes to reduce its packaging by 25%. For example, they are using less plastic in bottles of Liquid Tide detergent and Crisco Oil. They are using concentrates and combining laundry products into one package.

In Europe, P&G is selling small pouches of detergent that fit into a reusable package that people carry to the store and then back home to be used again and again. They are hoping Americans will be willing to do this, too, in the near future. They are also trying to find new ways to package their food products.

The PC Connection Company has cut down on packaging in many ways. One way is to have a large variety of different size boxes for shipping. They can fill each box up tightly and use only as much packaging as is needed for each order.

What You Can Do

With all the choices of packaging, and all the problems of making the best choice, you may wonder if there is anything you can do that will help. Here are some suggestions:

*Avoid red or yellow packaging. Toxic chemicals, cadmium and lead, are used to make the colors red and yelllow.
*Avoid "disposable" packages. They will just add quickly to the trash pile.

*Bring your own tote bag to the supermarket.
*Look for reusable or refillable packages that you can keep.
*Avoid products with too much packaging. For instance, buy loose apples rather than ones that are sealed in plastic.
*Avoid "blister packs" if you can get the same product unwrapped.
*Avoid products that have an unnecessary box holding a jar or can.
*Look for recycled and recyclable packages.
*If you buy something you use a lot, such as cereal or detergent, buy the largest box; avoid single serving containers that use the most packaging.
*write to companies that over-package their products; explain why they should change their packaging and suggest alternatives.

PLASTIC WRAPPER

MAINE AND THE STATE OF THE JUICE BOX

Amazing Invention, or Was It?

In the early 1980's a new type of packaging, called aseptic packaging, already in use in other countries, became widely available in the United States. It was thought to be highly innovative. It is light, unbreakable, safer than metal or glass, easily discarded, and, in the form of juice boxes, is easy to pack in a lunchbox. It doesn't need to be refrigerated. This packaging is referred to as aseptic because it protects its contents from bacteria that can cause disease. In locations where refrigeration is unavailable, particularly in under-developed countries, this is more than simply convenient. It means that milk and fresh juices can be made much more widely available to consumers.

The unique quality of this packaging is that it contains three materials bound together: paper, aluminum, and plastic. The aluminum foil forms a barrier against light and oxygen, preventing spoilage and changes in tastes. The plastic prevents spoilage by keeping out bacteria and prevents leaking. It also bonds the aluminum foil to the paper board. The paper board provides stiffness, strength and shape to the box. The boxes are usually sold in sets of three, shrink-wrapped, with a straw attached to each.

The aseptic package seemed to be a great idea. However, by 1989, four billion boxes were consumed and discarded. The state of Maine had set a goal to recycle one-half of its solid waste by 1994. Legislators were concerned that the non-recyclable juice box would interfere with this goal.

Maine Says No to the Juice Box

In 1989, Maine lawmakers put a ban on aseptic packaging. Juice boxes could no longer be sold in Maine. The ban was an expansion of the bottle bill enacted ten years earlier which required deposits on bottles, soda and beer cans. Due to the Maine law, the companies that produced the aseptic packaging began researching how to recycle the juice boxes in hopes that Maine would lift its ban.

Although no other state had a ban, sales of the juice boxes went down as consumers were concerned about the environmental impact of these non recyclable boxes. In at least one other state, Rhode Island, legislators enacted a bill that required aseptic packaging industries to work with the state to develop a pilot recycling program.

How is a Juice Box Recycled?

Currently, there is some juice box recycling, although there is much controversy about how environmentally sound this recycling is. The juice boxes are 70% paper, 23% plastic and 7% aluminum. In the recycling process, all three materials go through a hydra-pulper. The hydrapulper is like a large blender. (The process is similar to the process described in Activity 12, Making Paper.) The boxes are mixed in the blender with a lot of water. The paper becomes a pulp and sinks to the bottom. The plastic and aluminum float to the top and separate out.

The paper is made into paperboard and tissue and the plastic mixture is made into plastic lumber used for docks and porches. Plastic lumber is only produced in the Northeast. Recycling centers located outside the Northeast send the plastic mixture to landfills.

Is the Juice Box Environmentally Friendly or Not?

Those who argue that the juice box is environmentally sound might give the following examples to support their view:

1. The juice box saves energy. No refrigeration is required. Its light weight and its shape allow it to be stacked tightly. By taking up less space in transport, the amount of energy needed for transportation is reduced .

2. Plastic lumber is good because less toxic material is used to treat it than is used to treat wood.

3. It has less packaging than other forms of packaging when measured by weight. The aseptic juice box is 97% beverage and 4% packaging, as compared to glass bottles which deliver about 70% product to 30% packaging (all by weight).

Those who argue that the juice box is not environmentally

sound might question the viewpoint above in the following way.

1. How much energy is really saved by reducing refrigeration and transportation space? (The contents of most juice boxes don't need refrigeration anyway.)

2. There may be less packaging by weight but what about how recyclable it is? A lot of energy is needed to make the three layered material and a lot is needed to separate the three layers for recycling.

3. The same products could be packaged in glass or cans, both of which can be recycled more easily than aseptic packaging. Consumers can also package their own juice in thermoses.

4. Does it make economic sense? Recycling is only effective when some one will buy the product. Will people buy plastic lumber?

CHAPTER TWO

HAZARDOUS WASTE

OVERVIEW, BACKGROUND INFORMATION
AND SPECIAL PEDAGOGICAL CONCERNS FOR TEACHERS

OVERVIEW

The activities in Chapter Two offer students opportunities to learn about hazardous substances at home and in industry, experiment with source reduction, and learn about the disposal of hazardous waste.

In Activity 16, students identify products from their homes that contain hazardous substances; they also learn about alternative products or processes and try one alternative for a week.

In Activities 17 and 18, students learn about source reduction in manufacturing. Activity 17 is a simulation game. The students' goals is to make green squares with as little "hazardous" waste as possible. They learn that it is both difficult and possible. They also consider the role of source reduction on the trash issue and the responsibility of consumers on the manufacturing process. In Activity 18, students will see an example of a company that has reduced its hazardous waste.

Activity 19 gives students the chance to investigate hazardous waste in their schools and to suggest alternatives.

Through Activities 20 and 21, students will learn how controversial it is for our society to decide where to place a hazardous waste facility, and they will determine where the hazardous waste in their area is, actually taken for disposal.

BACKGROUND INFORMATION ON HAZARDOUS WASTE

The term "hazardous waste" is used to describe any substance that, when disposed of improperly, poses particularly dangerous threats to the health of the environment and humans. Such substances may be industrial by-products but may also be household products such as pesticides and cleaners.

Household hazardous wastes, when disposed of with the rest of the trash, may cause water pollution through leachate (toxic liquid that leaks out of landfills) or air pollution through emissions from resource recovery facilities (facilities that burn trash and generate electricity). Presently, there are no regulations about disposing of household hazardous waste. Attempts are being made to educate the public about such substances so people will separate them from the rest of their trash. Hazardous waste collection days are organized so unused hazardous products can be collected and disposed of in a hazardous waste disposal site.

Industries are now required to dispose of their hazardous waste following specific guidelines. The 1976 Resource Conservation and Recovery Act contains regulations for the safe disposal of hazardous waste and the monitoring of the disposal sites. The Comprehensive Environmental Response, Compensation, and Liability Act, also called the "Superfund," provides funds to clean up sites that pose immediate danger to surrounding communities. Nevertheless, monitoring the disposal of the huge volume of toxic waste currently produced in this country is daunting. Although new technologies and stricter regulations have made hazardous waste disposal safer than in times when such wastes were dumped in uncontrolled sites, it is difficult, if not impossible, to guarantee safety.

Because household hazardous waste is not regulated, it is impossible to determine exactly what percentage of this waste is from homes as compared with industry. However, it is widely thought that at least three quarters of hazardous waste is from commercial and industrial sources rather than household sources. It is for this reason that we have included activities and readings on industrial hazardous waste and waste reduction. It is also important to explain this to students for, while students can change their habits and make decisions to reduce their own production of hazardous waste, they should not be made to feel responsible for all hazardous waste. This pedagogical concern is discussed more fully below.

Methods of hazardous waste disposal include deep well injection (burying waste below ground water and aquifer levels in geologically stable formations), incineration and resource recovery facilities, landfilling, surface storage, and treatment plants. Each method poses its own particular potential hazards, and as citizens become concerned about possible hazards of such sites, communities resist the placement of sites close to their neighborhoods.

As our understanding of the health risks associated with toxic wastes grows, many people in government, industry, and public interest groups are calling for a concerted effort to implement source reduction, a strategy known as the 3 R's: reduction, reuse, recycling. Reduction of hazardous waste in industry means finding ways in manufacturing processes to use smaller amounts of toxic substances and to substitute less hazardous substances wherever possible. For individuals it means finding non-hazardous substitutes for those common household products that now contain hazardous substances. Reuse involves a company's recovery of a hazardous substance so it can be reused in the company's own manufacturing processes. Recycling of hazardous substances in industry means the use of one company's toxic by-product in another company's manufacturing processes.

Because the stakes are high for both the environment and for public health, many public interest groups have education programs to help the public become aware of the problem of hazardous waste disposal. A wide range of perspectives exists among government officials, industries, public interest groups, and concerned individuals about the best methods of disposal and regulation of hazardous waste.

SPECIAL PEDAGOGICAL CONCERNS: Questions and Suggestions for Teaching about Hazardous Waste

What are the issues that might come up when I teach about hazardous waste?

Hazardous waste disposal raises some of the most difficult issues in thinking and teaching about our waste problems. We learn that there are vast quantities of toxic substances produced daily and that it is difficult to dispose of such large quantities without threatening the health of the environment, including the humans that inhabit it. This is scary. Learning that toxic substances are not limited to the by-products of manufacturing but can be found in common products we use in our homes makes the problem all the more threatening. As time passes and the health hazards related to the disposal of various substances become more evident through research or, even worse, through contamination of the environment and illness of populations living near disposal sites, the need to find a solution to the problem becomes more acute. In addition, research in the last decade has shown that a disproportionate number of disposal sites for hazardous waste are located in low income and minority communities. The implications of this research are profound. Learning this information disturbs both our students and us.

How should I approach the issue of where hazardous waste disposal sites are located?

The social justice issues raised by the preponderance of hazardous waste disposal sites in poor and minority neighborhoods may be difficult for students to understand without a broader context in which to view prejudice and discrimination in our society. You may want to explore the general issue of discrimination with your students before turning to the material on disposal sites late in Chapter Two and throughout Chapter Three. Some good materials on teaching about these general issues can be found in the book *Open Minds to Equality* (Schneidewind and Davidson, 1983).

What should I do about my own strong feelings and those of the students as we study this information?

What do we, as teachers, do with our own feelings about the overwhelming nature of the problem and the injustice that has occurred because of the sitings? How do we guide our students as they become aware of, and then react to, the information? The answers are the same for studying hazardous waste as they are for studying other difficult contemporary issues. We cannot expect ourselves or our students not to feel anger, helplessness, fear and/or sorrow. In fact, it is important for them and for us to be able to express the feelings as they arise. For this reason, we suggest that journal writing be an integral part of the work in this and following sections. Allowing students time in class, even 10 minutes a day, to write their reactions to what they are learning allows them to give voice to their fears and anger. Sharing their responses in small groups is a good way to connect with others who have the same or different experiences of the information. It is important, as difficult as it is, not to try to deny or skip over the difficult feelings that arise from this study.

Is there any way to help students (and ourselves) feel hopeful rather than despairing about the problems related to hazardous waste?

We have designed the material on hazardous waste in Chapters 2, 3, and 4 in such a way that the information about hazards, and the questions of social justice that are raised by sitings, come in conjunction with information on alternatives, source reduction and taking action. In the following lessons students learn about hazardous waste in their homes and schools, and they learn about source reduction, one way to lower the amount of hazardous waste that ends up in the waste stream. We feel that the later lessons would give students an opportunity to see some hopeful resolutions to some of the problems and to take action themselves.

Taking action, or seeing that others are taking action, is an important step once the feelings of the students have been acknowledged. Students can research various groups that are helping to cut the amount of hazardous waste and assure more safe disposal methods. We have included some information about what is being done by various groups, individuals, and industries. Knowing that something is being done and, more important, doing something themselves can take students from despair to hope.

But what can students do about such a large problem? Educating others is one action that can make students feel hopeful. This can involve presenting what they know to other students, parents, or to the town through a letter or article in the newspaper. Auditing and reducing hazardous waste in the school is another way of making a difference that can be important to students. In the activity titled "Taking Action," we suggest a variety of projects that offer students ways to participate in working on the trash problem.

Finally, students need to know that although the information they are learning is distressing, knowing nothing makes us even less powerful. Only when we have the information can we work together to support groups and individuals who are working on solutions to the problem. The more people are aware of the problem, the more likely we are to find solutions.

Activity 16

HOUSEHOLD HAZARDOUS WASTE AUDIT

Objective

In this activity, students identify hazardous products in their homes and learn non-toxic alternatives to such commonly used substances. They will also become familiar with safe ways to dispose of household hazardous waste, having first learned about the effects of these materials in landfills and incinerators. In addition, students will experiment with ways of educating others about hazardous waste.

Note

Since students will be asked to survey hazardous waste in their own homes, it is important to obtain parental support for this project. The letter to parents in the introduction to this curriculum explains the purpose of the lessons on waste disposal, including the household trash audit and this household hazardous waste audit. Still, some parents may not want to participate. In this case, a student might find another adult, a teacher or neighbor, who would be willing to have a hazardous waste audit.

Students can identify hazardous products and their alternatives as an individual assignment, but in order to understand how these products become hazardous waste, they need to understand how landfills and waste-to-energy facilities work. We recommend earlier lessons in this book about landfills and waste-to-energy facilities or your explaining these concepts before doing this lesson.

We also suggest reading "Special Pedagogical Concerns: Questions and Suggestions for Teaching about Hazardous Waste," found in the introduction to Chapter Two. Since students can be disturbed to find themselves surrounded by potentially hazardous substances in their homes, we feel it is important to complete the second part of the assignment, in which students test and discover alternatives to hazardous products. This gives them the feeling that the situation is not hopeless and allows them to educate adults who also may not be aware of household hazardous substances and their alternatives. This knowledge will help students who may otherwise feel embarrassed by the amount of hazardous waste in their homes as compared to that of others. It is also important to explain to students that these household products generally become a problem *after* they are disposed of. When used with care, according to labeled instructions, they should pose no problem at home.

Suggested Previous Activities

  5. Constructing a Mini-Landfill
  7. Natural Cycles
  8. Burning Cans

Time

45 minutes to introduce activity
2-3 hours of homework
45 minutes for discussion

Vocabulary

caustic	landfill
corrosive	leachate
flammable	resource recovery
hazardous waste	toxic
incinerator	waste-to-energy
irritant	water table

Materials

1 toxic product brought in from home by each student
"Household Hazardous Waste Wheel," available from Environmental Hazards
    Management Institute, 10 Newmarket Road, P.O. Box 932, Durham, NH
    03824, or call them at 1-800-446-5256. It will be helpful to have at least
    one wheel per group of students.
1 copy of Handout 27, "Household Hazardous Waste Audit" for each student
1 copy of Handout 28, "Recommendations for Houshold Hazardous Waste" for
    each student
1 copy of Handout 29, "Do Alternative Products Work?" for each student
1 copy of "Battery Power: The Charge to Recharge" (part of Handout 29) for
    each student (for the suggested follow-up activity"

Supplementary Information for Teachers

A pamphlet from the Water Pollution Control Federation in Alexandria, Virginia gives a definition of a hazardous substance that is easy for young people to understand: "The Environmental Protection Agency considers a substance hazardous if it can catch fire, if it can react or explode when mixed with other substances, if it is corrosive, or if it is toxic." Many household products contain chemicals that, when improperly discarded, may contribute to the contamination of natural resources and water supplies. Contamination occurs through illegal dumping and leachate from landfills. It is at the time of disposal that hazardous substances become hazardous waste. Many household products are not harmful if used properly but still can become hazardous after disposal. Motor oil is a good example. It is necessary for running cars, but used oil contains benzene, which, when put in a landfill, can leach into water supplies or when incinerated, can produce toxic ash. To find out if a product is potentially hazardous, read the product label and look for words such as WARNING, POISON, CAUTION, HARMFUL, FLAMMABLE, EXPLOSIVE, IRRITANT, CAUSTIC, or HAZARDOUS. These words classify the type of danger possible from the product. The chart that accompanies this exercise gives examples of household toxic substances.

Activity Procedure

1. The day before you begin, ask students to bring in one item from home that they think would be a hazardous substance. Tell them what words to look for on labels. Teachers may also bring in such items as pesticides, bleach, bathroom cleaners, or motor oil as examples.

2. Explain that this work is an introduction to the household hazardous waste audit.

3. Have students discuss hazardous waste and where it comes from. (Many students associate hazardous waste with industries and most industries do produce hazardous waste.) Explain how household substances become hazardous waste (see background above). Although industrial hazardous waste disposal is now subject to regulation, there are currently no regulations on disposal of household hazardous waste.

4. Give each group of students a "Household Hazardous Waste Wheel."

5. In groups of three or four, have students read the labels of products they have brought from home. By using the wheel, students can determine:

   • if the item is toxic
   • the toxic chemical it contains
   • the proper disposal method
   • possible alternative products or processes.

6. Have the small groups report their findings to the class.

7. Encourage a class discussion to help students become aware of how useful and important some of these substances are to our modern life, yet how they pose a health threat at the same time

   * Why do manufacturers make products that contain hazardous substances?
   * Why do consumers buy them?
   * Do you think most people are aware of household hazardous waste?
   * What hazardous substances do you expect to find in your home?
   * What can people do to reduce the health threats posed by household hazardous waste?

8. Explain to students that they are going to do an audit of their house (or someone else's house) to find the products that could become hazardous waste. On the "Audit" (Handout 27) they should record: the name of the product, where they found it (location in the house), the toxic substance (found on the label or on the "Household Hazardous Waste Wheel"), and the classification (WARNING, POISON, FLAMMABLE, etc.). The classification identifies what type of danger is possible from the product. Give students several days to complete this assignment.

9. Once the students have completed the audit, they should complete Handout 28, "Recommendations for Household Hazardous Waste" either in class or as an additional homework assignment.

Each student should do an audit and recommendation sheet for his or her house, but students may want to work together, helping each other find the household hazardous substances and alternatives to them.

10. Have each student pick one item from the recommendation list, use that alternative product or technique for a week, and make an advertisement about it. Handout 29, "Do Alternative Products Work?" may be helpful. The article, "Battery Power: The Charge to Recharge" accompanies Handout 29 as an example of progress toward alternatives. The article may also be used as a follow-up activity as described below.

11. At the end of the week, when students have completed their advertisements, they can share the ads with the rest of the class. (If you have access to a video recorder, perhaps students could record their ads to share with other classes or the parent association.)

Suggested Follow-up Activity

Have students read "Battery Power: the Charge to Recharge." Hold a class discussion using the following questions:

* Have you used rechargeable batteries?
* What are the advantages and disadvantages of non-rechargeable batteries according to the article and your own experience?
* What are the advantages and disadvantages of rechargeable batteries according to the article and your own experience?
* How much money do you estimate you might spend each year to buy only non-rechargeable batteries?
* How much money can you estimate you would spend if you used only rechargeable batteries for one year?

This article and discussion offer an opportunity to discuss household needs and alternatives as well as introduce the economic questions and decisions involved in trying to reduce toxic waste. Activities 17 and 18 demonstrate industrial situations where people face similar questions.

HOUSEHOLD HAZARDOUS WASTE AUDIT

You are from the State Office of Waste Management. You have been asked to conduct an audit at the home of _____ to locate potential hazardous household substances and help the family use or dispose of them safely. You will complete the audit and then make some recommendations to the family about ways to decrease their use of products containing hazardous chemicals.

Product	Location in House	Toxic Chemical or Substance	Classification

RECOMMENDATIONS FOR
HOUSEHOLD HAZARDOUS WASTE

For the Family of _____

Prepared by _____

Below is a list of the hazardous materials in your household. For each hazardous product listed, I have suggested an alternative and/or a less hazardous product, and/or a technique to reduce the hazardous product's use, and a proper disposal method where appropriate.

Hazardous Product	Alternative Product	Reduction Technique	Disposal Method

Name _____

DO ALTERNATIVE PRODUCTS WORK?

Pick one alternative product listed on your Household Hazardous Waste Audit and try using it for a week. Make an advertisement about your experience. Your advertisement should include information on the hazardous product you were using, the alternative product you chose, and the disadvantages and advantages of each. Which product do you recommend consumers use, and why?

Write the script for your advertisement below.

BATTERY POWER: THE CHARGE TO RECHARGE

We live in a battery-powered world: from Walkmans to Gameboys: from cameras to watches: from the memories that store numbers in our phones to the powerpacks that charge our cordless tools—batteries are almost everywhere. Unfortunately, although these batteries make our lives more convenient, they're not without their drawbacks. In order to fuel the estimated 900 million battery-operated toys and appliances now in use, Americans go through 2.5 billion batteries a year. And many of those batteries end up in local landfills, where they have the potential to

leach dangerous chemicals into the environment.

However, the good news on batteries is that this situation is about to change. Across the country, the recycling movement is growing: manufacturers are beginning to introduce batteries made with lower levels of toxic chemicals; and rechargeable batteries are steadily gaining in popularity.

A GOOD GROUNDING

All dry-cell batteries create electrical power via a chemical reaction in them between two metals. Various battery types rely on different combinations of metals. Some of these chemical reactions allow you to recharge a battery: others do not.

Primary and secondary batteries are the two types you're likely to use in your home. Primary batter-

ies are the traditional single-use cells. Once their energy supply is depleted, they cannot be used again. There are several types of primary batteries: alkaline, one of the most popular kinds: carbon zinc, the batteries used before alkalines were introduced: mercury or mercuric oxide, small "button" batteries used primarily in watches and hearing aids: and silver or silver oxide, another type of button battery.

Secondary batteries are rechargeable. Although there are several types of secondary batteries, the only ones you're likely to buy for the toys and gadgets in your home are nickel cadmium. Other secondary batteries include sealed lead acid, found in camcorders, tools, emergency lighting, and other battery-powered products; nickel metal hydride, used in laptop computers; lithium for computers; and zinc air, a relative newcomer that relies partially on air as a catalyst to fuel the unit's chemical reactions.

By far, most of the batteries used in homes are not rechargeable. Alkaline and carbon zinc batteries make up more than 50 percent of the battery market—and account for a large portion of the 2 pounds of batteries per household per year that are thrown out. In addition to the waste that these batteries create, throwaways are also an environ-

mental headache because both alkalines and carbon zincs contain mercury, a toxic metal that can damage the brain, nervous system, liver, kidneys, and also can cause genetic disorders.

There are currently no recycling programs for alkalines and carbon zincs in the U.S. However, because of mercury's potential danger if leached into landfills, environmental groups have begun to call for such programs. So far, the battery manufacturers have resisted, arguing that the process of extracting mercury from these batteries is too expensive and that there is no viable market for the reclaimed metals. "They can't purify the materials enough for them to be used again by the industry," says Lisa Silverstone, associate staff executive for the National Electrical Manufacturers Association. "There are some places in Europe, however, that do this processing, but it costs $3,000 a ton to effectively recycle alkaline batteries."

Instead, the battery industry has responded to environmental concerns by producing alkalines with no added mercury. (Mercury was originally added to batteries to increase their running time.) These batteries, which contain only minute amounts of mercury (Sanyo's is 99.9995 percent mercury-free), will soon be the only alkalines available. The new batteries, according to the manufacturers, mean that there is no need to recycle alkalines. No-added-mercury batteries can safely go into landfills because they contain mostly inert metals.

Others disagree, however. "People with a more environmental bent are worried about placing concentrated groups of any type of battery in land-fills, especially in older landfills without linings," says Judith Kornberg, waste management program director for the Environmental Action Coalition. Even the no-added-mercury batteries, in large enough numbers, could pose a threat, she says. "We worry that if the outer casings corrode, there might be some leaching of mercury."

So far, there's no real solution in sight. Several states, including New York and New Jersey, have formed task forces to examine the options. Under New York law, for example, batteries cannot be incinerated. Burning batteries can release mercury into the air. The task force is now studying ways to keep batteries out of both incinerators and landfills. One option, for instance, would be to bury batteries only in "secure" landfills lined with layers of plastic, sand, and clay. But, says Kornberg, "There's no real way to recycle them at the moment."

POSITIVES OUTWEIGH THE NEGATIVES

In the meantime, environmentalists urge the use of rechargeable batteries when possible. "Anything to reduce the number of batteries that are thrown away each year is preferable," says Kornberg. And, rechargeables are more powerful and convenient than ever.

"There are a lot of outdated and negative perceptions about rechargeables," says Timothy Hill, marketing and sales director for Millennium Power Systems, which supplies brand-name rechargeables to other battery companies and manufactures its own line of rechargeables. "These are a legacy of the technology's nascent stage—when the batteries didn't run for a long period of time and when it took 17 hours to recharge them."

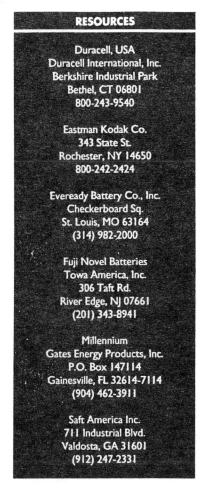

RESOURCES

Duracell, USA
Duracell International, Inc.
Berkshire Industrial Park
Bethel, CT 06801
800-243-9540

Eastman Kodak Co.
343 State St.
Rochester, NY 14650
800-242-2424

Eveready Battery Co., Inc.
Checkerboard Sq.
St. Louis, MO 63164
(314) 982-2000

Fuji Novel Batteries
Towa America, Inc.
306 Taft Rd.
River Edge, NJ 07661
(201) 343-8941

Millennium
Gates Energy Products, Inc.
P.O. Box 147114
Gainesville, FL 32614-7114
(904) 462-3911

Saft America Inc.
711 Industrial Blvd.
Valdosta, GA 31601
(912) 247-2331

BATTERY GUIDELINES . . .

TYPE	USE
Carbon zinc	Household appliances, toys, etc.
Alkaline	Household appliances, toys, etc.
Mercury (battery manufacturers are phasing these out due to mercury content)	Hearing aids, medical devices, pagers, button batteries for cameras, watches
Silver	Button batteries for watches, cameras, etc.
Nickel cadmium	Rechargeable; used in tools, toys, etc.
Lithium	Cameras, computers, button batteries

* Chart derived from information provided by the Mercury Refining Company, Inc.

Today's nickel cadmium batteries are easier than ever to recharge. Instead of 17 hours, it takes from one to five hours to repower a battery, depending on the brand, type of battery, and the recharging unit. The recharging units are more compact than they used to be, too. Millennium, for example, makes "Charge Man," which is about the size of a small electric razor. Designed for portability, it will repower two AA cells in three to five hours. Today's chargers also turn off when the batteries are fully-powered, allowing you to store the batteries in the plugged-in charger so fresh batteries are always on hand.

Typically, one rechargeable battery will take approximately 1,000 charges, meaning that it replaces about 300 alkaline batteries. Besides cutting down on waste, you'll save money and energy. It takes about one penny of electricity to charge a nickel cadmium battery, says Hill. In other words, it costs less to recharge a battery than to burn a light bulb for the same amount of time, he says. One study at Carnegie-Mellon University found that over three years, using rechargeables in a personal stereo saved more than $600, including the cost of the charger and the cost of the electricity to power the batteries. (Rechargeables run about $6 to $8 for a two-pack; chargers sell for around $20.) The rechargeables also reduced the amount of solid and toxic wastes created by up to 85 percent, according to the Carnegie-Mellon study.

Rechargeables are not a cure-all. Alkaline batteries are a better choice in some situations where long-lasting power is necessary, such as in smoke detectors. That's because nickel cadmiums will lose their power faster than alkalines, which have a longer shelf life. Nickel cadmiums run for about one-third to one-half the time that an alkaline will in the same application; then they must be recharged.

The other drawback to rechargeables is the toxic metals they contain. Cadmium has been linked to kidney and respiratory cancer, and nickel can destroy nasal tissues and cause asthmatic lung disease.

However, unlike alkalines, nickel cadmium batteries can be recycled once they can no longer take a charge. In fact, several manufacturers, including Sanyo and Millennium, encourage consumers to mail their used nickel cadmium batteries back for recycling. Sanyo even sells its rechargeable batteries in a plastic tube that can be used as a mailing container.

The final problem with rechargeable batteries has to do with their use in cordless appliances: That cordless screwdriver or vacuum cleaner you keep plugged into the outlet actually has a nickel cadmium battery pack in it. Once the batteries can no longer take a charge, you have to throw the tool out—sending it, and the toxic metals it contains, to the landfill. By 1993, however, you will be able to replace those hidden batteries. New "easily removable" laws in some states mean that manufacturers of cordless appliances will be forced to make the batteries simple for the consumer to remove and replace.

Although the "easily removable" laws will help keep some batteries out of landfills, manufacturers worry about consumers accidentally putting throwaway batteries into appliances meant to take only rechargeable—a dangerous mistake. Attempting to recharge a single-use battery can cause it to explode. Battery manufacturers are working on a solution to the problem, though. Millennium, for example, has come up with an alkaline battery feature called Intellilink. Should you accidentally put alkalines instead of rechargeables into a cordless appliance, Intellilink will "recognize" the batteries and not attempt to recharge them. Result: The screwdriver will not be damaged nor the alkalines explode.

Until the technology to recycle throwaway batteries is realized, environmental advocates recommend using rechargeables as much as possible (and, when possible, recycling used rechargeables). If you use throwaways, call your local municipality to check about collection and disposal regulations. Regulations and collection programs vary widely from city to city.

by Devra Pine

BATTERY DOs AND DON'Ts

• Read instructions carefully. Use the size and type of battery recommended by the manufacturer of your battery-powered appliance. Be sure to insert batteries in equipment according to the manufacturer's instructions.

• Store batteries in a dry place at room temperature. Batteries do not need to be refrigerated to stay fresh.

• Never carry batteries loose in your pocket or purse, where they can come in contact with metal objects such as loose change. The batteries could short out and leak.

• Remove batteries from equipment that won't be used for several months because they can leak.

• Remove batteries from equipment and appliances before you power them with household (AC) current. Otherwise, they could explode.

• Replace all the batteries in an appliance at one time; mixing old and new batteries together can cause them to leak. In addition, replace worn out batteries as soon as possible to prevent them from leaking on equipment.

• Never throw batteries into a fire. They can explode.

• Keep battery-powered equipment away from extreme heat or cold. The expansion or contraction can cause the batteries' cases to crack and leak.

• Never attempt to recharge a disposable battery such as an alkaline or carbon zinc battery. They can explode. Recharge only batteries that are marked "rechargeable."

• Keep batteries out of reach of small children.

Activity 17

GREEN SQUARE GAME: INDUSTRIAL HAZARDOUS WASTE

Objective

By simulating the operations of a manufacturer, students will experience the problems of making a product without generating hazardous waste. They will consider the responsibilities of industries and consumers in the production of toxic substances. Students will also learn how source reduction relates to hazardous waste. They will understand that hazardous waste, even when properly disposed of, can still be a threat to the environment.

Suggested Previous Activity

16. Household Hazardous Waste Audit

Time

45 minutes

Vocabulary

hazardous waste
source reduction
toxic

Materials

1 4" x 4" piece of heavy grade paper painted green
Each group of 3-4 students needs:
4 cups
2 brushes
1/2 cup of water in a full cup container
blue tempera powdered paint
yellow tempera powered paint
2 spoons
1 4" x 4" piece of heavy grade paper
table covering (newspaper)

Supplementary Information for Teachers

All manufacturing produces waste. Much manufacturing produces waste that can be harmful to humans, animals, and/or the environment. Such waste is disposed of either in waste-to-energy or resource recovery facilities, incinerators, or landfills. Incinerators, while destroying a great deal of the toxic chemicals through burning, release air pollution and produce highly toxic ash as a by-product. Ash is then placed in landfills, where rainwater percolating through the waste can carry it to ground water and aquifers, particularly in older landfills built before requirements for clay or plastic liners were instituted. Some toxic waste can be rendered non-hazardous by treating the waste with chemicals that neutralize the toxic substance. However, toxic waste, once produced, is difficult to get rid of.

Toxic use reduction is the practice by which industries change their processes and/or materials used in production in order to decrease the amount and toxicity of the waste produced. Since industries must pay to dispose of their toxic waste, toxic use reduction often makes the production process more economical.

Activity Procedure

1. Before class, mix some green paint and paint a 4" x 4" piece of heavy grade paper. This green square is the model of the customer's desired color, to be left on display during the activity.

2. Divide students into groups of three or four. Explain that each group is a company that manufactures green squares. Encourage groups to come up with company names.

3. You, the teacher, are a customer who wants to purchase green squares. You have prepared a model of the color green you are looking for. You will choose the company you buy from based on how accurately the color is reproduced and how little hazardous waste is produced in the process. (You are an environmentally conscious consumer!)

4. Explain that any surface or object that becomes contaminated with paint, whether blue, green, or yellow, becomes hazardous. This includes all materials (brushes, cups, spoons, etc.), hands, clothing, the table surface and the floor. Mention that the backs of the squares should not be contaminated with paint either. Write the list of criteria for your purchasing decision (listed below) on the chalk board for students to see.

5. Allocate roughly 20 minutes for the painting activity. Suggest that one group member record the steps taken by the company to arrive at the color green.

6. When groups have painted their sample green squares, it is time for the customer to decide which company to buy from. The customer evaluates the green square companies using the criteria listed below: On the chalk board, draw a chart listing each company's name (student group names) and the criteria for selection with possible points assigned to each category. Add up the scores to find who will win the contract. You will be counting the number of <u>uncontaminated</u> items, so the points are always positive points.

    Criteria for decision:
    • color match with model - 10 points (Note that the color has to be close or the customer may not consider awarding the contract, despite good environmental conservation; teachers may assign each company points ranging from 0 - 10.)
    • production floor cleanliness - 1 point
    • number of uncontaminated brushes - 1 point each
    • number of uncontaminated cups - 1 point each
    • number of uncontaminated spoons - 1 point each
    • clear, uncontaminated left over water - 1 point
    • cleanliness of back of green square -1 point
    • no leftover green paint - 1 point

- uncontaminated table covering - 1 point
- uncontaminated hands, clothing, etc. - 1 point for each clean student

7. Give points for each area and add up the points to select the company.

8. Hold a class discussion using the following questions about waste issues. You may also want students to reflect on the decision-making process that each group experienced.

* What techniques did you invent in your atempt to minimize waste?
* Which techniques reduced waste? Which ones did not?
* What methods did you try in order to attain the desired color?
* Did these techniques work?
* Did any people have ideas that they did not say but wish they had?
* How might this game relate to the operations of real manufacturers?
* Does concentrating on generating less waste make manufacturing easier or more difficult, or does it have no effect?
* Do you think it costs more or less to make a green square when you pay attention to source reduction?
* What responsibilities do you think manufacturers have in disposing of their hazardous wastes?
* What responsibilities do consumers have? In this game the consumer weighted the color by giving it 10 points. Will consumers take their second choice for something if they know less waste was produced in the process?
* What items do students use or wear that were manufactured by industries that may have produced hazardous waste? (almost everything)
* How would your group make decisions differently if given the chance? Should we try it again?

9. Continue the discussion in order to understand the shuffling of waste that currently takes place when it leaves such a company.

* What does happen to waste generated by the green square factory?
* What do you think happens to the water used to clean off the brushes and other items used in the production process? (The water used to clean manufacturing equipment gets contaminated with waste and is sent to a waste water treatment facility.)
* What happens at the waste water treatment facility? (The toxic metals are removed in the treatment process.)
* What happens to the heavy metals? (They are sent to a hazardous waste incinerator.)
* Then what happens? (Bottom ash, air pollution, and fly ash -- toxic ash that gets caught in the pollution control equipment of the incinerator -- are the byproducts.)
* What happens to the toxic ash? (It is put in a landfill.)
* What happens when water seeps through the landfill? (The water becomes contaminated with the toxic chemicals. If possible, this leachate is collected, turned into a solid form, and incinerated, leading to another round of toxic ash.)

The cycle continues. (See below the example of this cycle, which could be drawn on the chalk board.) While the volume of the waste is reduced, the resulting waste's toxicity may be higher than the original waste and the waste is never completely eliminated. How should we deal with hazardous waste generated in industrial processes? Is source reduction an answer?

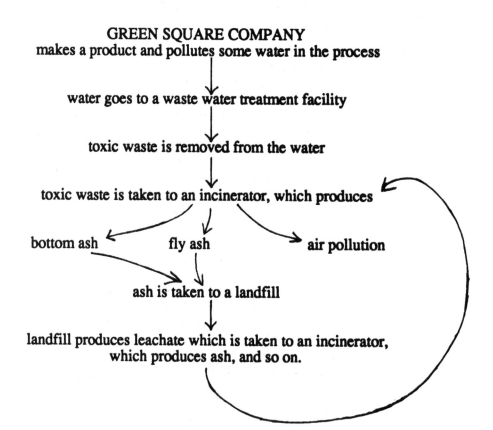

GREEN SQUARE COMPANY
makes a product and pollutes some water in the process

water goes to a waste water treatment facility

toxic waste is removed from the water

toxic waste is taken to an incinerator, which produces

bottom ash   fly ash   air pollution

ash is taken to a landfill

landfill produces leachate which is taken to an incinerator, which produces ash, and so on.

Activity 18

AN EXAMPLE FROM INDUSTRY: 3M'S 3P'S

Objective

Students will learn about the 3M Company's success in decreasing hazardous waste production. Students will find out what companies in their area are doing about source reduction.

Suggested Previous Activities

13. Source Reduction: Brown Bag Lunch
17. Green Square Game: Industrial Hazardous Waste

Time

45 minutes to introduce activity
homework (telephone interview)
45 minutes for information sharing

Vocabulary

hazardous waste
recycling
source reduction

Materials

1 copy of Handout 30 "3M's 3P's" for each student
paper and pencils
phone numbers of local industries and state environmental offices

Activity Procedure

1. Make sure the students understand the term "source reduction". Go over this term if they have not done previous activities relating to source reduction.

2. Read the handout on 3M's 3P's.

3. In small groups, have students make a list of how 3M reduced the amount of hazardous waste it produced.

4. Have each group report the list to the class and go over how source reduction is accomplished. (See "Background Information on Hazardous Waste" in the introduction to Chapter Two for further information about source reduction.)

5. With the whole class, design a questionnaire that students can use to interview local industries.

6. Assign each small group or individual students to interview a local industry. They may visit the plant or call on the phone. Go over interviewing etiquette: students should identify themselves and their purpose, make an appointment, and tell the person how long the interview will take.

7. Give students a few days to complete the interview.

8. Have students share their information with the class. (Perhaps the class would like to write letters commending companies that are working on source reduction and recommending it to those who are not.)

Variations on the Activity

A. One group of students can interview a state official instead of an industry representative. They can report what the state is doing about source reduction of hazardous waste in industries to the class.

B. The EPA has a listing called a "toxic release inventory," which contains the names and addresses of companies that use or emit chemicals that are defined as hazardous. If the company uses over 10,000 pounds or emits over 25,000 pounds of a toxic chemical, they must report this to the EPA. Such information is available to the public. You can call the EPA and inquire about particular towns. (See a listing of all EPA regional offices in the resource section of this curriculum.) They will tell you the companies in those towns that are using or emitting toxic chemicals. Students can use this information to write letters to these companies, inquiring about their use and emission of toxic chemicals and whether the company has considered implementing source reduction. You and your students can also contact your regional EPA office to find out if there are any suspected or known hazardous waste sites in their area. To do this, call the regional EPA office and ask for the Superfund Community Relations Department. They should be able to provide you with a complete listing.

3M's 3P's

One manufacturer that has been working on pollution prevention since 1975 is the 3M Company. In that time they have made changes in their manufacturing processes that have resulted in both the elimination of pollution and the reduction of costs. According to an article published in the *Environmental Professional* (1989) by two members of 3M's Environmental Engineering and Pollution Control, the program "Pollution Prevention Pays," ("3P") was started for practical and philosophical reasons.

In the 1970's, as environmental awareness grew, laws and regulations about pollution became more strict. Companies were required to make sure that their emissions of potentially hazardous substances did not pose a threat to people or the environment. As a result, many companies started treating such substances, but only after they had been produced. These sorts of solutions -- scrubbers in smokestacks to reduce air pollution, water treatment facilities to clean contaminated liquid before it leaves the plant -- are called "end-of-the-pipe" solutions. End-of-the-pipe means something that is done once a pollutant has been produced.

Generally, these solutions are expensive and can cost companies a lot of money. As a result, many companies such as 3M are trying to reduce the amount of hazardous substances they produce and use in order to have less pollution that will eventually need to be treated. So there was a practical, money-saving reason for 3M to look for ways to reduce pollution. This practical attitude became a company policy with the following commitments:

- to solve its own environmental pollution and conservation problems;
- to prevent pollution at the source wherever and whenever possible;
- to develop products that will have a minimum effect on the environment;
- to conserve natural resources through the use of reclamation and other appropriate methods;
- to assure that its facilities and products meet and sustain the regulations of all federal, state, and local environmental agencies;
- to assist, wherever possible, governmental agencies and other official organizations engaged in environmental activities.

These commitments were introduced to employees at all levels of the company, and employees were rewarded for discovering ways to reduce pollution. One idea was to use new and different resources that would create less pollution. A second idea involved using a different process that produced less pollution. A third idea required that the company change its equipment design in order to perform better and more efficiently. Fourth, resource recovery was initiated, which means reusing the by-products of a given process in other processes, or selling these by-products to other companies to be used in their processes.

Employees or departments that discovered ways to reduce pollution were recognized publicly and rewarded with dinners and special outings. New ideas were published in short one-page papers like the one about cleaning copper that follows.

3M estimates that since 1975 when the program began, they have reduced pollution by 450,000 tons and have saved nearly $420,000. 3M is an example of how companies that make a commitment to reducing pollution can make significant changes and save money for the company in the long run, as well as help keep the environment clean.

Pollution Prevention Pays

■ *Managing for a better environment*

A Compendium of 3P Success Stories

From Shower to Scour

Waste Stopper: Pumice on Copper

Problem

3M's electronic products plant in Columbia, Mo., makes flexible electronic circuits from copper sheeting. Before sheeting can be used in the production process, it has to be cleaned.

Formerly, the metal was sprayed with ammonium persulfate, phosphoric acid and sulfuric acid. This created a hazardous waste that required special handling and disposal.

This 48-inch diameter pipe carries solvent-laden air from plant, right, to be used as supplemental boiler fuel in adjacent building.

Solution

Cleaning by chemical spraying was replaced by a specially designed new machine with rotating brushes that scrubbed the copper with pumice.

The fine abrasive pumice material leaves a sludge that is not hazardous and can be disposed of in a conventional sanitary landfill.

Payoff

- 40,000 pounds a year of hazardous waste liquid prevented.

- $15,000 first year savings in raw materials and in disposal and labor costs.

- In the third year of use the new cleaning machine had saved enough to recover the $59,000 it cost. Because of increased production each year, costs saved and volumes of pollution prevented continue to rise.

The Idea Team

The technical employees who conceived and implemented the new cleaning method:

William V. Ballard,
Manufacturing, Columbia, Mo.

Dwaine L. Haagenstad,
Division Engineering,
St. Paul, Mn.

Fred J. Luden,
Manufacturing, Columbia, Mo.

Werner Wiedemann,
Laboratory, St. Paul, Mn.

3M

Activity 19

HAZARDOUS WASTE IN SCHOOLS

Objective

By becoming "toxic use detectives," students will understand how their school contributes to the hazardous waste stream and discover what they can do to decrease this problem.

Note

This lesson is divided into two sections. Part A gives details for source reduction in chemistry labs. Part B emphasizes other items in the school such as cleaning materials (see Household Hazardous Waste Wheel for alternatives, or products distributed by Seventh Generation or Earth Care Paper, both listed in this curriculum's resource directory), office supplies (Wirth Co., Inc., Hayward, CA 94545 produces a non-toxic liquid paper), art department materials (see *Health Hazards Manual for Artists*, Michael McCann, Nick Lyons Books, NY, 1985), and pesticides used in the school yard and in the building (see Household Hazardous Waste Wheel).

Suggested Previous Activities

16. Household Hazardous Waste Audit
17. Green Square Game: Industrial Hazardous Waste
18. An Example from Industry: 3M's 3P's

Part A: Source Reduction in Chemistry Labs

Time

45 minutes

Vocabulary

hazardous waste
source reduction

Materials

paper and pencils
any guides or information obtained from sources listed below
multiple copies of information gathered from school's chemistry department
1 copy of Handout 31, "Hazardous Waste in School," for each student or pair of
    students

Supplementary Information for Teachers

In chemistry labs, there are ways to reduce the amount of chemicals produced (source reduction), and there are specific disposal methods for hazardous chemicals. Three approaches to look for are: (1.) decreasing the amount of each chemical used, (2.) substituting non-hazardous chemicals for hazardous ones, and (3.) treating hazardous chemicals so they are no longer hazardous.

Microscale is a technique used to decrease the amount of hazardous chemicals produced in a chemistry lab. If your school is already using Microscale, perhaps the chemistry teacher would be willing to explain to the class how hazardous waste is reduced by the process. If your school is not using Microscale, information on its techniques and other ways to deal with hazardous waste can be gained from the following sources:

Reduction of Hazardous Waste from High School Chemistry Labs
Dr. Wahl Jr., Editor
Department of Chemistry
North Carolina State University
Raleigh, NC 27695-8204
tel: 919-515-2941
$3.00

Microscale Organic Lab, 2nd Edition (1989)
D.W. Mayo, R.M. Pike, S.S. Butcher
John Wiley and Sons
New York, NY

Hazardous Chemicals Disposal and Information Guide
M.A. Armour, L.M. Browne, G.L. Weir
CRC Press. 2000 Corp. Blvd. NW
Boca Raton, FL. 33431
(407) 994-0555

Waste Disposal in Academic Institutions (1990)
James Kaufman
Lewis Publishers
121 South Main St.
Chelsea, MI 48118
tel: 1-800-272-7737
$55.00

Destruction of Hazardous Chemicals in the Lab (1990)
George Lunn and Eric B. Sansone
John Wiley and Sons
New York, NY
$65.00

Environmental Hazards in Your School: A Resource Handbook
Available free of charge from the EPA
Publication #2DT-2001 (October 1990)

Contact the Environmental Protection Agency or your local state environmental agencies for regulations on disposal of hazardous chemicals and for additional information.

Part A - Activity Procedure

1. You or a student should obtain from the chemistry/science department in your school a list of all the hazardous chemicals that are used, how they are used, and how they are disposed of. Make multiple copies of this list, as students will need to consult this information early in their research.

2. Have students work in pairs or individually. Assign each person or pair one chemical to research. (Use information from one of the guides.)

3. Have students answer the questions on the Handoout 31 worksheet

4. Hold a class discussion where students report their findings and suggestions for source reduction of hazardous substances to other students. Discuss whether chemistry can be taught without creating hazardous waste.

5. As a class, draft a letter to the science department, principal, superintendent, or board of education expressing your concerns and suggesting alternatives you have discovered.

HAZARDOUS WASTE IN SCHOOL

1. What is the name of the chemical you are researching?

2. What are some of the dangers associated with this chemical?

3. How is the chemical used in your school?

4. How does the school dispose of this chemical?

5. Is this disposal method the best one? Why or why not?

6. Is there another chemical that could be used instead of this toxic one?

7. What are your recommendations to the school for dealing with this chemical?

8. What are the sources of your information about this chemical?

Part B: Hazardous Waste In Schools

Time

Advance preparation: Send for catalogues from Earth Care Paper and Seventh Generation. (See Resource Directory for addresses.)
45 minutes to begin activity
1 hour for homework (reading and research)
45 minutes class for follow-up

Vocabulary

hazardous waste	source reduction
herbicide	toxin
pesticide	

Materials

1 copy of Handout 32, "School Decides to Monitor Pests Rather Than Clobber Them," for each student
paper and pencils
catalogues from Earth Care Paper and Seventh Generation
1 copy of Handout 31, "Hazardous Waste in School," for each student or pair of students

Activity Procedure

1. Divide students into pairs. Tell them they are going to be toxic use detectives.

2. Assign students to various locations in the school (office, lunchroom, classroom, art room, custodian's office, photography room, etc.). It may be helpful to alert other staff that students will be doing this activity. Conversely, any difficulty that students encounter in getting information can be a useful teaching opportunity.

3. Have students make a list of all the items that they think could become a hazardous waste when thrown away. (One area they may overlook is pesticides and herbicides that may be used on athletic fields and in the building. The principal or business manager should have this information.)

4. Discuss student findings. Did they discover many toxins? In what area were the most found?

5. If students are unsure about which items are toxins, have them contact their local EPA office or refer to the Household Hazardous Waste Wheel.

6. What can be done? Have students read the Handout 32 article on pesticide use. Through telephone calls, letters, or personal visits, have them research the policies of other schools in adjoining areas. Compare these to their findings from their own school.

7. Discuss source reduction. In pairs or individually, have students pick one chemical from their list to research. Guides suggested in Part 1 may be helpful, as well as the Earthcare and Seventh Generation catalogues. Students should answer questions on the "Hazardous Waste in School" handout.

8. Encourage a class discussion with the following questions:

* What did you find?
* Are some chemicals necessary? Which ones?
* Are some chemicals replaceable? How?

9. Students may wish to write letters to various teachers, department heads, or administrators to give suggestions on source reduction. Also, students can make posters to advertise their findings.

School decides to monitor pests rather than clobber them

THE BOSTON SUNDAY GLOBE • JUNE 23, 1991

By Caroline L. Cole
SPECIAL TO THE GLOBE

LEXINGTON – When insect exterminator Dick Pingree shows up in the Lexington High School kitchen to take care of stray cockroaches or ants, he leaves his can of Dursban pesticide in his truck.

Instead, he takes a clipboard and pen so that he can record evidence of any insect infestations that may have occurred since his last monthly visit. Should one of his "sitcky traps," located strategically under sinks and in dark corners, contain the telltale evidence of an unwanted pest, Pingree must first notify John Moynihan, the school's supervisor of buildings and grounds before he can use the can of pesticide.

According to a Boston Globe survey of the 25 school systems in the NorthWest Weekly area, Lexington is the first school system to adopt a formal policy against general pesticide usage to control unwanted insects and rodents. Initiated last September, the aggressive inspection program is being heralded as a model by parents, school administrators and even Pingree's employer, Waltham Chemical Co.

"The fact is we don't want to see one roach in our kitchens and there haven't been any," said Moynihan. "If one of the inspectors were to notice a problem, we would first find how the insects are getting in. We'd patch the crack or put in a new screen before we would request spraying."

Like many schools in the region, Lexington had had a long-standing annual contract with Waltham Chemical Co. to do preventive spraying in the kitchens, cafeteria, teacher's room and food storeroom. The School Co. mittee agreed to ban regular pesticide applications after the Lexington League of Women Voters alerted administrators to the inherent dangers in general pesticide usage.

"Dursban and Diazinon, two pesticides Waltham Chemical was using in the schools, are neurotoxins," said Myla Kabat-Zinn, the parent who chaired the League's study of indoor-air pollution. "We were concerned about their use in the schools without notification to parents because some people are more sensitive than others to chemicals. We also concluded the insects could be controlled more effectively using nonchemical methods."

Kabat-Zinn points out that regular applications of pesticides without concern for whether pests are present, get at the symptom of infestation without addressing the problem and might even set the stage for a future infestation. "We discovered regular spraying renders the chemicals useless over time because the insects begin to develop a tolerance," Kabat-Zinn said. "So you have to use larger doses or switch brands for the chemicals to work."

Philip Geiger, Lexington's superintendent of schools, enthusiastically embraced the idea of a nonchemical program in part because he discovered pesticides had been making some students sick. "Since we began notifying parents on days when we are forced to spray, we've had a handful of parents with legitimate concerns who have kept their kids out of school, particularly those children with asthma," he said. "The inspection program costs slightly more, but the cost is insignificant if you consider the discomfort to some students."

Schools can legally use pesticides to control insects and rodents as long as they are applied according to their label instructions, said Gail Kaprielian, chief of the Pesticide Bureau at the state Department of Food and Agriculture. "The pesticides used must be registered by the state, and the applicator must have a license issued by this department," Kaprielian said.

But Kaprielian agreed with Kabat-Zinn that the state's regulation of how, where and when pesticides are applied is "very, very broad." Kaprielian said, "The label might say a particular pesticide cannot be applied in a manner that would knowingly cause harm, but determining whether or not the intent of the regulation has been violated can be difficult since adverse health effects may not be easily identified."

To limit risk to the public, Kaprielian's department is in the process of updating its advisory on the use of pesticides in occupied rooms. "We are particularly concerned with the use of pesticides around children because they are often on the ground near where the pesticides are being applied and can come into contact with the chemicals more readily," Kaprielian said.

But Kabat-Zinn said such efforts to regulate the application of pesticides give her little comfort. "The state may say these chemicals are OK today, but we haven't been using them long enough to really know what damage they may cause to future generations," she said.

"Are counters washed down after the pesticides are applied? Are the rooms properly vented when the exterminator is working? Are efforts being made to keep the chemicals from coming in contact with pots and pans?" asked Kabat-Zinn, listing some of her concerns.

"In some schools, custodians are regularly called upon to apply pesticides to take care of such acute problems as wasps and bees. Are they following the label directions completely? I bet that, as was the case in Lexington, most schools have long-standing contracts that nobody is questioning or even thinking about. The exterminator is just part of the landscape," Kabat-Zinn said.

The Globe's survey of pest control in the region's schools confirmed Kabat-Zinn's assertion that administrators and school food service directors are often unaware of pesticide use in the kitchens. In more than one case, the buildings and grounds supervisors said no pesticides were being used in their school kitchens. But food service directors contradicted them by producing bills for exterminating services. In other schools, the food service director was uncertain when the exterminator came, what chemicals were being applied and what should be done to limit risk to staff and students.

Kabat-Zinn said that cockroaches and mice are indeed harmful, but, she added, "since everyone, even the chemical companies, acknowledge pesticides are toxins and may cause cancer, why not err on the side of safety and try other methods first?"

Donald Rivard, district manager at Waltham Chemical, said he welcomed the opportunity to apply new methods of pest control in Lexington, but said his company has run into resistance from school staff. "Custodians will complain we're not doing the job because they don't see us doing any spraying," Rivard said. He said the monitoring program costs about $200 more annually, or $3,400 a year, than the pesticide-based program because of the extra administration involved.

Moynihan believes he has addressed the educational problem with his custodial staff by advising them early in the year of the benefits of the nonchemical method. "The custodians were also involved installing screens and plugging cracks and crevices to keep the bugs out," he said.

WHAT OTHER COMMUNITIES ARE DOING WITH PESTICIDES

Here is how other NorthWest Weekly schools are handling pest control in their facilities:

Acton: Monthly pesticide application by professional exterminator in food service areas, according to Mary Branon, food service director.

Andover: Monthly pesticide application by professional exterminator in food service areas, according to Frank Paul, facilities manager.

Arlington: Monthly pesticide application by professional exterminator in food service areas, according to Pat Isai, food service director. Extermination firm has begun more monitoring procedures.

Bedford: Monthly pesticide application by professional exterminator in food service areas, according to Lawrence Canter, business manager.

Belmont: Monthly pesticide application by professional exterminator in food service areas, according to Robert Martin, supervisor of buildings and grounds.

Billerica: Monthly pesticide application by professional exterminator in food service areas, according to Jane Ford, food service director.

Burlington: Information was not made available.

Carlisle: No regular pesticide applications.

Chelmsford: Monthly pesticide application by professional exterminator in food service areas, according to Lorraine Small, food service director.

Concord: Monthly pesticide application by professional exterminator in food service areas, according to William Hutchinson, manager of maintenance and custodians. Custodians can use chemical spray bombs on wasps and bee infestations.

Dracut: Information was not made available.

Lawrence: Information was not made avialable.

Lowell: Exterminator is on an as-needed contract and is called in as frequently as twice a month, particularly to kill roaches in the swimming pool and locker rooms, according to Fred Buckley, senior custodian.

Medford: A professional exterminator is called when needed to control pests in classrooms and nonfood areas, according to William Jenkins, supervisor of buildings and ground. Jenkins ordered spraying for termites at the Lincoln Junior High School last summer. Republic Management, the school's food service contractor, would not release information about pesticide use in the kitchen areas referring the call to Jenkins. Jenkins said the kitchens and cafeterias were not under his jurisdiction.

Methuen: Regular pesticide application by professional exterminator in food service areas, according to Alan Alvino, supervisor of buildings and grounds. Mouse traps are set out in winter.

North Andover: Regular pesticide application by professional exterminator in food service areas, according to Maria Messina, food service director.

North Reading: No regular pesticide applications in food service area, according to food service director Jean Pietrillo, because no insect problem has been noticed.

Reading: Monthly or bimonthly pesticide applications by professional exterminator in food service areas, according to Catherine Cook, food service director. Door sills and hallways at the high school, middle school, Birch Meadows and Barrows elementary schools are treated quarterly for termites.

Stoneham: Termite treatment in one classroom at the Robin Hood Elementary School and monthly pesticide application at the middle school.

Tewksbury: Exterminator is called in on an as-needed basis, according to Mickey Sullivan, director of facilities, athletics and maintenance. Each of three schools has had one treatment this past school year.

Westford: Information was not made available.

Wilmington: No regular pesticide applications. Professional exterminator would be called if custodians spot a problem, according to Roger Lessard, custodial supervisor.

Winchester: Monthly pesticide application by a professional exterminator in food service areas, according to William Miller, food service director.

Woburn: Regular pesticide application by professional exterminator in food service areas, according to Thomas Quinn, head of maintenance services. Custodians use over-the-counter chemicals on wasps and bees. Some ant traps used.

Activity 20

HAZARDOUS WASTE INCINERATOR SITING

Objective

In this activity, role-playing will help students understand the various perspectives of interest groups and individuals in support of and opposed to the siting of a hazardous waste incinerator. Students will practice decision-making skills in the process.

Suggested Previous Activities

8. Burning Cans
10. Do You Know Where Your Trash Is?
11. What Should Be Done with All the Trash?
17. Green Square Game: Industrial Hazardous Waste
18. An Example from Industry: 3M's 3P's

Time

1 or 2 45 minute periods

Vocabulary

incinerator   resource recovery
hazardous waste   waste-to-energy

Materials

1 copy of Handout 33, "Background Information for Role Play" and packet of
letters for each student

Activity Procedure

1. Explain that students will be doing a role play about the siting of a hazardous waste incinerator proposed in their town. The students will present a variety of perspectives to the site selection committee which will make a recommendation about the use of the site.

2. Divide students into small groups of three or four. Assign each group one of the following roles:

- the governor's site selection committee
- citizen living in the town of proposed site
- citizen not living in the town of proposed site
- local newspaper reporter
- industry owner

• local engineer

3. All students should read Handout 33, "Background Information for Role Play," and each group should locate the letter from the packet that they feel supports their position.

4. Each group should prepare an argument for their position, either opposing the site or supporting it, to present to the site selection committee

5. Each group should choose one person to present their argument to the site selection committee.

6. The site selection committee will call on the spokesperson to represent the position to the committee.

7. Site selection members should listen to presenters and write down ways they agree or disagree with the arguments presented. They should also write any questions that concern them.

8. After all presentations have been made, individuals may ask questions of other groups. The site selection committee will monitor this discussion and designate a time period during which it must take place.

9. After interested groups and individuals have had a chance to ask questions, the site selection committee may ask questions.

10. The committee should confer and make a recommendation at the end. (Note: The recommendation does not have to be for or against the incinerator. The recommendation could be that the governor needs more information before making a decision.)

11. Hold a class discussion using the following questions:

* Given that the state produces a certain amount of hazardous waste that must go somewhere, where should it go?
* Who will be affected by the disposal site?
* What responsibilities do citizens of one town have to citizens of another town when a disposal site is chosen?
* What if the site chosen is a rural community, where one-fourth the number of people will be affected as compared to an urban site? What responsibilities do citizens from other towns have in that case?
* If the site selection committee doesn't make a recommendation, what impact will that have on the disposal of waste now being produced?

BACKGROUND INFORMATION
FOR ROLE PLAY

An incinerator company, Clean Air, has proposed to build your state's first hazardous waste incinerator. Your state has prospered in recent years and has become the fourth largest generator of hazardous waste in the United States. Presently 250,000 tons of hazardous waste are shipped to other states each year.

Since your state does not have facilities to dispose of its hazardous waste, it must export the waste. The state presently exports a greater share of its hazardous waste than any other state in the nation. The federal government has mandated that all states that generate hazardous waste must have "capacity assurance," which means they must be able to dispose of the waste produced in their own states. Congress may soon ban the interstate transportation of hazardous waste. Also, fees to dispose of the waste in other states are rising rapidly.

There is concern from the residents of Anytown (where the incinerator will be sited) that the incinerator will pose health hazards. The Health Department in Anytown has previously noted a higher than normal percentage of respiratory problems due to pollution from the shipyard that used to be located there. The town would probably receive 55% of the emissions from the incinerator's stacks. There is also concern that the 3.7 million pounds of pesticides that would be transported to the incinerator each year pose the threat of an accidental spill. (3.7 million pounds. of pesticide weighs the same as 82 Statues of Liberty put together.)

The incinerator company, Clean Air, submitted a 7,000 page report showing the incinerator would have almost no environmental impact. They argue that closeness to the incinerator would not be a health hazard because the incinerator will be safe. Clean Air calculated that in the plant's projected 20-year life span, it would cause less than one case of cancer.

To: Editor, Anytown News

I am not a citizen of Anytown; I live in a little village a few miles south of Anytown. I have been following the debate about whether to allow Clean Air to build a hazardous waste incinerator. Part of the right of being a citizen of the United States is to speak up, even when you may live in the town down the road, as I do.

Specifically, I think there are a lot of merits in Clean Air's proposal. I am worried that opponents in Anytown will not even consider the advantages of the proposal. It seems that tempers are flaring on both sides of the issue, and in the end Clean Air may just get tired of waiting and take its business elsewhere. Where will that leave us? Just with the hazardous waste and nothing else.

To those citizens in Anytown who are protesting the Clean Air proposal, I suggest that you think about it a little more. The state is suddenly in terrible economic condition right now, and we may have to take the risk of building a hazardous waste incinerator in *your* city so that everyone in the state can have jobs and pay their bills. And you are not the only ones having to take a risk: I know that the trucks carrying hazardous wastes to the incinerator will pass within 300 yards of my home. I'm not happy about that, but I am not a selfish person who only thinks of myself and ignores everyone else.

If you are against the building of the incinerator, please see if you can do some investigation; find out how to do it better; help both sides come to a compromise; help educate people about the issues. Maybe you can come up with an idea that will be better, but don't block Clean Air if you can't. The rest of the state will be watching you.

Your fellow citizen and neighbor,
Susan Snyder

To: Editor, Anytown News
From: Robert L. Bishop
President, Clean Air Enterprises

As President of Clean Air Enterprises, I have been involved in every aspect of the proposal to build one of our modern and ecologically safe incinerators in your fine community. We have been impressed with the warm welcome your citizens have given to us on all of our visits. I want to commend even those of you who are opposed to the plan, because you have been polite and thoughtful in all of our meetings. I want to use this letter to talk to those of you who are opposed to the plan, and remind you of the important role a hazardous waste incinerator can play in the growth of Anytown, the state, and this region of our country.

First, the federal government, under the Superfund Act, requires the state to have a plan as to how they will dispose of all the hazardous wastes its industries generate by the year 2000. Your state is the fourth largest hazardous waste producing state in the country. We are here to handle that hazardous waste, cleanly. If the state fails to do so, the state will lose money, first from the federal government, and then it will lose jobs, lots of jobs, as industries go to other states. The three years it takes to build one of our modern facilities will not only bring in valuable construction jobs, but will allow the state to handle its toxic waste well before the deadline.

Our hazardous waste incinerator will make the region safer for the present and future. Hazardous wastes are piling up, and many can no longer legally be placed in landfills -- our hazardous waste incinerator will safely eliminate the toxins before they do harm. Right now trucks are traveling through the entire state, carrying hazardous wastes hundreds of miles to neighboring states -- a locally located facility would let us dispose of the wastes Anytown creates more quickly, safely and inexpensively. And what will happen when nearby states pass laws banning the import of hazardous wastes from your state? Even the federal government is considering banning the exporting of wastes to other countries. How big a mountain of hazardous wastes can Anytown and this state safely contain without a hazardous waste incinerator?

Finally, there will be more jobs in Anytown, and for many years to come. In addition the town will be generating taxable revenues from income and the state's transportation tax.

Clean Air Enterprises is proud to be doing our part in making the environment cleaner and safer, and in restoring economic pride to the state. We truly hope you will soon be able to reach agreement on our proposal, and allow us to begin construction before we are involved in other projects in other areas. Thank you again for all your hospitality.

To: Editor, Anytown News
From: The United Industry and Business Council

The United Industry and Business Council (UIBC) supports the hazardous waste incinerator proposal submitted by Clean Air Enterprises. The Council is especially concerned that the region's future economic growth and the availability of federal Superfund assistance may be in jeopardy if this facility is not approved. This is the seventh time such a facility has been proposed, and it is time we got the job done.

We say this with a certain degree of caution. The Council has always been in the forefront of environmental quality concerns, and we know that no hazardous waste incinerator can be absolutely guaranteed to be perfect. But we simply cannot afford to wait for perfection. Already we are shipping 30% of all waste generated in the region to other states, at the cost of $13 million. There is a strong possibility that this option will soon be prohibited. If so, and without our own hazardous waste incinerator, we can easily predict the results:

1) Many of our current industries and businesses will be forced to relocate to other regions. The loss of income will prevent the state from having enough money to build a hazardous waste incinerator and the region will die out. No industry that generates any hazardous waste will come here again. This includes almost all industries.

2) The state will lose its Superfund money.

3) The existing hazardous wastes will become a dangerous source of pollution that cannot be moved.

4) Clean Air Enterprises may soon be hired by another region to build a hazardous waste incinerator, and we will have to start from scratch if we are ever to get this project underway in time.

Already we are losing jobs. Recently the hazardous waste incinerator many of us use in a nearby state was shut down for three months to upgrade its pollution control equipment. We simply had to shut down our own factories also, since we had no other options for our hazardous wastes. It resulted in a temporary loss of a number of jobs. The Council has made a priority of reducing our hazardous wastes, but it is impossible to eliminate them completely. We must develop a local way to handle them.

If those who are opposed to the Clean Air proposal can come up with a better way, please make it known. The need for a hazardous waste facility will not disappear even if Clean Air's proposal is rejected. Right now the future of our region is at stake, and time is running out.

To: Editor, Anytown News

I am just a citizen of AnyTown, and have been my whole life. I've brought up my family here, and hope to spend the rest of my life here. It is a wonderful place to live.

But now I am worried about the future. The proposal to build a hazardous waste incinerator in our town worries me. I never thought one town should be the place where everybody gets to bring their wastes. It seems to me it might be better if each factory was responsible for what it produces, and then we could spread the hazards around. I certainly don't dump my garbage in my neighbor's yard. Heck, he'd probably just throw it right back.

But this is the real world, and maybe we folks in Anytown have no choice. I'd hate to see the local industries move elsewhere -- I have three grandchildren who would probably lose their jobs. And if we do get the hazardous waste incinerator, there'll be jobs in the area for years to come.

I understand that there are sometimes dangers with these facilities. They don't always work right, they don't get rid of all the hazards, and there's a chance there'll be some air pollution problems, but can we escape air pollution anyway with all the traffic today? The folks who oppose the plan just don't seem to trust what the CleanAir people say. But I think the people in Anytown have had a good experience with big business, and I'm willing to give them another shot.

Yours truly,
John Martin

To: Editor, Anytown News

As parents who chose to move to Anytown because we were so impressed with its record for public safety and democracy, we are firmly opposed to the building of the hazardous waste incinerator by Clean Air Enterprises.

If you look at a map of the town, you will see that over 30,000 people live within one mile of the proposed site, and there are 200,000 people within a 5 mile radius. That includes a tremendous number of children who will have to breathe in all the pollutants the incinerator spits out each day. Remember, even the people who support these facilities admit that some toxic fumes escape. How many toxic fumes do you want your children to be breathing? We believe you would say, "None!"

We remember the 1970's, when the oil companies were bragging about how safely they would be shipping oil from Alaska. They had all that money, all that technology, and they would be doing everything right, and they said it was for our own good. Then there was the Valdez disaster.

In India there was the horrible explosion at the Bhopal plant, killing thousands. Then we had the Three Mile Island nuclear power disaster. In Russia, they experienced the melt-down at the Chernobyl plant, which spread nuclear wastes over thousands of miles and destroyed the way of life of entire regions of the country.

In each case the people who built these facilities assured the public that the risks were insignificant. We believe they really tried to avoid disasters. And each case involved complex, enormous projects, involving potentially dangerous materials. We can't allow Anytown and all of its citizens to be put in similar risk.

Clean Air Enterprises wants to burn up to <u>five tons of hazardous wastes per hour</u>. Trucks bringing hazardous wastes would travel down our main streets, passing residential areas, schools, business districts, and the town reservoir. Also, in case of disaster, it would be very difficult to evacuate residents on these same congested roads.

Clean Air Enterprises' safety record is not good. With NO experience in operating a hazardous waste incinerator, Clean Air has had their current operations fined and shut down innumerable times due to toxic spills, traffic violations, operating without a license, discharging improper materials into the sewage system, and many other violations.

Yes, we do need a hazardous waste facility. But not, absolutely not, in such a congested place as Anytown. Why can't we find a place that is less crowded, or use isolated state land, or part of a shut-down military facility.

It is up to all of us -- parents, grandparents, business people, teachers, students -- to make our town safe from disaster. Let us not gamble with our lives and land in a rush to build a hazardous waste incinerator in our town.

Yours truly,
Lester and Beatrice Washington

To: Editor, Anytown News
From: The Anytown Site Selection
Committee

We would like to thank all of the citizens of Anytown who have supported us in our efforts to make the facts known about the proposed hazardous waste incinerator. We would also like to report that we have had pleasant meetings with the representatives of Clean Air Enterprises, who would like to build the facility, although they have yet to provide answers to the many questions we have about the safety hazards of their proposed site in Anytown. It is of utmost importance that the citizens know the facts. Without them, we may put ourselves in grave danger:

1) Hazardous waste incinerators pollute the air. The closest house to the site is only 350 yards away, and 200,000 people live within 5 miles of the site. The American Lung Association issued a report in January 1990 saying that the health costs of air pollution are now $40-50 billion. CBS news reports that the death rate from asthma has increased and environmental pollution is a factor.

Clean Air's facility will be burning benzene. This is a cancer causing toxin. They will also be burning waste oil and cancer-causing PCBs.

2) Our region is already plagued by air pollution from our industries. Nearby Centerville has 55% more lung cancer than is normal. We can't afford to pour any more toxins into our air.

3) No hazardous waste incinerator works perfectly. Even the best equipment, according to Clean Air's own numbers, destroys 99.99% of toxins. What they don't destroy-- and remember, they are planning to burn at a rate of 5 tons per hour--will go into our air and into our lungs. And when equipment wears out and operators make mistakes, we will breathe in the results.

4) Horrible mistakes happen. Remember Three Mile Island?

These are the facts that Clean Air Enterprises and their supporters won't tell you. Please consider these facts in making your decision. From the standpoint of public health and safety, a hazardous waste incinerator does not belong in a densely populated area such as Anytown.

To: Editor, Anytown News
From: Beyond Toxic Waste, Inc.

In the debate about the possibility of placing a hazardous waste incinerator in Anytown, we believe that no one has mentioned any alternatives. At Beyond Toxic Waste, Inc., we work with industries to reduce and eliminate their toxic waste stream, and ultimately do away with the need for hazardous waste incinerators.

Dealing with our nation's waste volume is not simply a matter of finding somewhere to put it or burn it. We can, if we are committed, to never produce it in the first place. Companies such as DuPont, IBM, Dow Chemical, and Hewlett Packard are trying the reduction approach, and discovering that it pays off in a short time. This effort can be expanded. If we follow this path, we can stop investing our money now for incinerators that may soon be useless.

Government must give industry a clear signal that waste reduction is a priority. Here's how:

1) Companies can change the raw materials they use in production. They can change the production technology and equipment used.

2) Wastes can be recycled in the factories.

3) The actual products themselves can be redesigned to eliminate the need for toxic wastes.

4) Companies can exchange their waste products -- "one person's waste is another's treasure."

5) Waste management practices can be improved.

6) Toxic wastes can be neutralized, not burned into the air.

This change can be accelerated if citizens refuse to buy products that contribute to toxic waste pollution. Industry can be encouraged to do a better job - if it has enough reasons.

We are opposed to building any hazardous waste incinerators. It makes no sense in any long-term plan for the environment, and instead is a serious problem of its own. These hazardous waste incinerators add to pollution and don't encourage us to cut down on toxic wastes.

A recent development is Westinghouse's portable, high-temp incinerator designed for "on-site" use. At least this approach would not put any one community at great risk. By working on different ways to dispose of toxins we will not be creating giant incinerators that can do great harm.

Beyond Toxic Waste, Inc. is committed to a cleaner future. We say we can do better than shift the pollution from landfills and water into the air we breathe. Rather than building monstrous incinerators, we can start building a better future today.

Activity 21

WHERE DOES HAZARDOUS WASTE
FROM YOUR STATE GO?

Objective

By contacting state environmental protection agencies and companies that handle hazardous waste, students will find out what happens to the hazardous waste their state generates.

Note

The research in this activity is a homework assignment that includes phone calls. The phone calls may be frustrating at times because often it takes awhile to locate the person with the correct information. You may want to contact the people your students will be calling and ask their permission to have students call them. In class, students will present their findings. Before giving students this assignment, you will need to have a list of local businesses and companies that are likely to handle hazardous waste. Such businesses might include dry cleaners, gas stations, and furniture refinishers. Just about every manufacturer generates hazardous waste, so include on your list any local industries, such as auto, food, jewelry, clothing, or shoe manufacturers etc. Perhaps the town clerk can give you a list of local manufacturers.

It is possible that students within the class may live near a hazardous waste facility and may feel anger or embarrassment as this information is realized. You can use activities in Chapter Four (Making a Difference) to increase these students' sense of empowerment and help them make connections with other students in the class.

Suggested Previous Activities

16. Household Hazardous Waste Audit
17. Green Square Game: Industrial Hazardous Waste

Time

20 minutes to introduce activity
1-2 weeks for homework assignment (It often takes a while to reach the
        appropriate person on the phone.)
25 minutes for follow-up discussion

Vocabulary

hazardous waste

Materials

list of local businesses and industries that handle state-wide hazardous waste
several copies of Handout 34, "Where Does Hazardous Waste Go?,"for the 3
students who will call the EPA
1 copy of Handout 35, "How Companies Handle Hazardous Waste," for each of
the other students.

Activity Procedure

1. Ask the class where hazardous waste comes from. (primarily from industry, but also from households)

2. Ask students what happens to the hazardous waste produced in their state. Explain that their homework assignment will be to answer this question.

3. Three students will contact the state department of environmental protection or the EPA (Environmental Protection Agency) office that covers your region to find out where the hazardous waste in the state is disposed of; they will record their findings and conclusions on Handout 34. The other students will contact companies in the state to find out how they dispose of their hazardous waste and what cost is involved; these students will complete Handout 35. Discuss the telephone etiquette of students identifying themselves and their purpose when contacting an agency or company.

4. Have students report their findings to the class. Discuss the questions below.

* What were some interesting facts about the amount of hazardous waste your state produces?
* What kinds of actions do companies take to dispose of hazardous waste? Do these actions cost much?
* Do you think your state has found a good way to deal with hazardous waste?
* Do you think anyone would want a hazardous waste facility in his or her town?
* Under what conditions would you want a hazardous waste facility in your town?
* What do you think states should do with the hazardous waste they produce?
* Do you think each state should be responsible for its own waste? Should the federal government be involved? Should each company be responsible?
* What is your proposal for dealing with hazardous waste?

Name _____

WHERE DOES HAZARDOUS WASTE GO?

Find out what happens to hazardous waste in your state. Start by calling your state office of environmental protection or management or the EPA (Environmental Protection Agency) office that covers your region. Report all your findings below. You may get different information from different people. Keep track of whom you talk to and what information you get from them. Fill out a new worksheet for each person you speak to. When you have finally gotten all the information you need and can fully describe where the hazardous waste in your state goes, answer question #2.

1. a. With whom did you talk?

   b. Where does this person work?

   c. What is his/her title?

   d. What information did you obtain about where the hazardous waste in your state goes?

   e. How does the amount of hazardous waste your state produces compare to that of other states?

2. From the information you collected, what conclusions can you make? Was it difficult to obtain this information? Did you have to talk with many people? Why do you think this was difficult, or why was it not difficult?

Name _____

HOW COMPANIES HANDLE
HAZARDOUS WASTE

Contact a company in your area or state. Find out how much hazardous waste they produce and how they get rid of it. Fill in the information below.

1. What company did you contact, and where is it located?

2. What kind of products does it produce?

3. With whom did you talk? What was his/her title?

4. What hazardous waste does the company produce, and in what amounts?

5. How and where do they dispose of it?

6. How much does it cost to dispose of the waste?

7. From the information you collected, what conclusions can you make? Was it difficult to obtain this information? Did you have to talk with many people? Why do you think this was difficult, or why was it not dificult?

CHAPTER THREE

HAZARDOUS WASTE AS A SOCIAL JUSTICE ISSUE

OVERVIEW AND BACKGROUND INFORMATION FOR TEACHERS

OVERVIEW

Chapter Three starts with Activity 22, a small group and whole class discussion on prejudice, discrimination, racism, and classism. We encourage teachers to use this activity or others serving a similar purpose to prepare students for later activities.

Activity 23, "In My Backyard?," is a game where students have to decide on a site for a hazardous waste incinerator. The game is very simple in structure, designed to focus on the human decision-making process among people faced with hard choices, rather than on the technical or scientific issues as in Activity 20, "Hazardous Waste Incinerator Siting."

Activities 24 and 25 offer case studies of environmental injustice. In Activity 24, students will take roles in a skit of a meeting between Navajos and waste facility representatives. They will also be able to read articles to find out what decision the Navajos made.

Activity 25 includes short summaries of two historically important hazardous waste landfill sites. We selected these two because they were among those first studied by the General Accounting Office (GAO). The Warren County, North Carolina site, which prompted the GAO study, is now closed. The Emelle, Alabama site is still functioning and is the largest hazardous waste landfill in the world. Those two summaries give students some background as to how the questions about environmental racism became public.

BACKGROUND INFORMATION ON SOCIAL JUSTICE ISSUES

During the last decade, research has shown that our waste, especially our most dangerous waste, is dumped disproportionately in areas populated by poor and racial/ethnic minority people. The explanation for this is difficult and complex and must be seen in the larger context of our political, economic, and social structures. The term "environmental racism" has been used in recent years to describe the decisions and effects of this disproportionate dumping. Throughout the following activities, encourage students to consider whether or not this term is accurate by posing the following suggested questions:

* What makes a decision or situation "racist"?
* How is this different from a responsible or fair decision or situation?
* What does it mean to belong to a "class"? What does it mean to be classist?

* Are a person's race/ethnicity and class related?
* Do you think government and corporate decisions about hazardous waste siting could be viewed as racist or classist? Why or why not?
* If so, what should/can they do now?
* Do you think the government or these corporations thought they were making a responsible decision? Why or why not? How does this affect your judgment of their actions?

The information about the demographics of communities surrounding hazardous waste disposal sites received widespread public attention as a result of the protests against a proposed location for a hazardous waste disposal site in Warren County, North Carolina. The protesters charged that selection of the site was motivated by racism, and when the lawsuits failed, two U.S. representatives requested that the General Accounting Office (GAO) study the correlations between hazardous waste disposal sites and the racial and economic demographics of surrounding communities in the southeastern United States.

The GAO study (GAO, 1983) looked at four hazardous waste landfills in eight southeastern states. (Although this study focused on southeastern states, this does not mean that dump sites are not located in other geographic regions of the United States.) These sites were chosen because they were the only offsite landfills (that is, those not adjacent to an industrial facility) in that area. The report, published in June 1983, revealed some astounding findings. Of the four sites, three had populations that were more than 50% African American; at the fourth site, African Americans represented 38% of the surrounding population. In all four areas, African Americans represented at least 90%, and in two cases 100%, of the families that lived below the poverty level. When the GAO.report was published, some people saw the findings as clear evidence of racism and classism in the decision-making about toxic waste disposal.

A few years later, a broader study of demographics in communities where hazardous waste disposal sites are located was completed by the Commission for Racial Justice of the United Church of Christ. Their map of the United States, documenting minority population and disposal sites, appears in Activity 25. The Commission's report, completed in 1987, concluded that "the findings of the descriptive study on the location of uncontrolled toxic waste sites suggest an inordinate concentration of such sites in Black and Hispanic communities, particularly in urban areas."

The statistical data shows clearly that poor and minority communities bear the burden of housing toxic waste disposal sites. The Commission on Racial Justice further concluded that "the possibility that these patterns resulted by chance is virtually impossible."

When asked to explain the findings, Dr. J. William Porter, assistant administrator for hazardous waste for the U.S. Environmental Protection Agency (EPA) said that the EPA did not decide where sites should be located. State governments are responsible for choosing sites, while the EPA is responsible only for approving them. He said, "We primarily look at technical factors, emissions, ground water monitoring. Sociological factors are not part of our review." This response raises the question of who should be reviewing the sociological factors.

If states are responsible for selecting disposal sites, what factors do they take into account? Are their decisions based on demographics, and if so, are these decisions racist or classist? One example may serve to demonstrate. In 1984, the California Waste Management Board was having difficulty deciding on a disposal site. They hired a consulting firm, Cerrell Associates, to analyze demographics for possible resistance to a site for a waste-to-energy facility meant to handle the growing waste problem in Los Angeles. The study, using a variety of factors, labeled which communities would strongly resist such a siting and which would not. Those that would resist were liberal, college educated, young or middle-aged, middle- or high-income groups in urban areas. Those least likely to resist were "communities that conform to some kind of economic need criteria," particularly "lower socio-economic areas" or "heavy industrial areas, with little, if any, commercial activity."

At a time when the public awareness about the potential dangers of transporting and disposing of hazardous waste is increasing, more and more communities are resisting sitings in their neighborhoods. In an attempt to find disposal space, decision-makers may solve the problem by looking for the area of least resistance among their constituencies. If governments make decisions about waste disposal sites based on which communities are least likely to resist, what will the results be? Because the reasons for the higher representation of minority groups in this country's poor neighborhoods reflect the continuing problems of racism in our society, we believe that the study of hazardous waste and racism/classism must be seen in the larger societal context.

Activity 22

LEARNING ABOUT RACISM AND CLASSISM

Objective

Students will consider and discuss the terms: prejudice, discrimination, stereotype, racism and classism. They will draft definitions of these words to be revised throughout the chapter.

Note

The activities which follow in this chapter introduce students to environmental racism and classism regarding our nation's trash disposal policies. Activity 22 offers the opportunity for students to think about these terms unto themselves in preparation for applying them to the issue of trash production and disposal. While there are many forms of systematic discrimination against groups of people other than racism and classism, these are the relevant terms regarding environmental decision-making and effects, and thus are the only forms addressed here.

Time

45 minutes

Vocabulary

classism	racism
discrimination	stereotype
prejudice	

Materials

paper
pencils

Activity Procedure

1. Divide students into groups of 3 or 4.

2. Have students choose one person to record answers and another to be the spokesperson for the group.

3. Ask the students to discuss and try to define the following words in this order: prejudice, stereotype, discrimination, and then racism and classism. Teachers may want to have dictionaries, social studies books that deal with these topics, or other resources

available. Encourage students to use examples from history, literature, their own lives, and so on, to identify these terms.

4. Discuss the definitions with the whole class. Ask students about the words in the same order as above: prejudice, stereotype, discrimination, racism and classism. It will be important for later activities for students to begin to distinguish their meanings.

We have included some definitions below as *supplementary information for teachers*. Most importantly, help students understand the difference between opinions and feelings based on insufficient grounds, and having the power to act on and reinforce those opinions and feelings.

Encourage students to revise their definitions as they progress through the other Chapter Three activities. They may be able to add examples or depth to their initial definitions.

There are many ways in which this exercise could be expanded: students could interview people who work on topics related to these terms, guest speakers could be invited to class for discussions with students, students could read newspapers to locate examples of prejudice, racism, and so on.

Supplementary information for teachers

Prejudice is an attitude, feeling, or opinion formed without sufficient knowledge or reason. It is a judgment made prematurely, a predisposition. While a prejudice can be a positive or negative judgment, it often carries a negative connotation, as in a prejudiced person (someone who is not open to new ideas) or a prejudicial statement (a comment meant to influence).

A **stereotype** is a generalization about the characteristics of a group of people. Stereotypes are based on an oversimplified view of a group of people and are assumed to be inherent in all members of that group without distinguishing any individual's unique characteristics. Stereotypes can appear to be favorable or unfavorable, but in either case they restrict individuals and do not allow a deeper understanding of that group of people.

Discrimination is the act or practice of according differential treatment to one group of people over another based on group identity, including class, race, religion, gender, sexual orientation, language, etc. Discrimination is an action, not an attitude or feeling, like a prejudice or a stereotype.

Racism and **classism** are prejudices, stereotypes, and discrimination that are systematically enforced. The suffix, *ism,* denotes the ideology or conditions in which discrimination based on race or class exists. Racism and classism involve an exercise of power which carries out differential treatment of one group over another. Racism and classism are reinforced by institutional practices, historical legacy, and cultural attitudes and values. (Cross-Cultural Consultation, 1992).

Class is a socio-economic category. One's socio-economic class is determined by numerous factors including: actual wealth (e.g. money and property), access to wealth (e.g. one's relatives and friends), access to resources and institutions (e.g. physical and mental health care resources and institutions), level of education and status of educational credentials, status of profession (related to actual salary or not).

Racism and classism operate on various levels. Some researchers describe these levels as attitudinal, cultural and institutional. Others use the categories of individual, interpersonal, institutional, and ideological. Whatever the categories are, they describe the many levels of human interaction, and the many means by which people define what characteristics and behaviors are the norm by which others are judged.

Activity 23

IN MY BACKYARD?

Objective

Through a voting process activity, students will become aware of some of the problems of deciding on a site for a hazardous waste incinerator. They will also be introduced to the question of inequality in the placement of such a facility.

Note

This activity is a simulation. Students will have to decide where to place a hazardous waste incinerator. In Round #1, students will know where they live and will have equal voting power. In Round #2, students will live in the same places as in Round #1, but "power cards" will give them unequal voting strength. This is different from the role-play on selecting a site done in Activity 20, where students considered technical, geographic, and other information. "In My Backyard?" has a very simple structure, making the only focus the interaction between people.

In real life, individual citizens each have one vote when they go to a voting booth. However, citizens don't get to vote directly on the placement of an incinerator. Instead, as students learned in Chapter Two, recommendations are made to a state's governor and legislature; then those state officials decide on a location. State officials are influenced by many people and many factors. The difference in voting power in Round #2 represents different levels of influence that citizens hold. Help students understand this and encourage them to think about what contributes to divergent levels of influence.

It is possible that students within the class may live near a hazardous waste facility and may feel upset or self-conscious as this information is realized. You can use activities in Chapter Four (Making a Difference) to increase their sense of empowerment and help them make connections with other students in the class.

Suggested Previous Activities

    11. What Should Be Done with All the Trash?
    20. Hazardous Waste Incinerator Siting
    21. Where Does Hazardous Waste from Your State Go?

Time

    30 - 45 minutes

Vocabulary

hazardous waste	toxic
incinerator	incinerator
landfill	

Materials (per group of 6 or 7)

      1 copy of Handout 36, "State Map"
      1 copy of Handout 37, "Fact Sheet"
      1 set of town cards
      1 set of power cards
      1 copy of Handouts 38, "Round #1 Worksheet" and Handout 39, "Round #2
          Worksheet"

Activity Procedure

1. Divide students into groups of six or seven

2. Give each group a map, a set of town cards, and a fact sheet (Handout 36, 37). There should be one student per town. Give the appropriate map -- the six-town map or the seven-town map -- or adjust one of them.

3. The map should be placed in the middle of the group. Each player is dealt one town card, which indicates where he or she lives.

4. Instruct players that they are members of a state with six (or seven) towns. Further instruct them that a new federal law mandates that all states must have a hazardous waste incinerator to deal with their hazardous waste and that it must be located in their state. They **cannot** ship their hazardous waste elsewhere. The citizens must decide where to locate this facility. Encourage them to portray their towns as they wish; they can give their towns characteristics related to terrain, population, industry, and so on.

5. Read the fact sheet out loud.

6. Appoint a note-taker for each group.

7. Give students three minutes to decide where the facility will go. At the end of three minutes, the note-taker should read the questions on the "Round #1 Worksheet", (Handout 38). The group should discuss the questions, and the note-taker will write down their responses.

8. Now give each group a set of power cards. They should keep the same map and town card as in Round #1. Each student is dealt a power card. This card indicates the number of votes that each student/citizen now has (people no longer have equal votes). Once again, the goal is for the town to decide where to put the facility. Allow five minutes for strategizing before the vote is taken. Students may realize that they can lobby other towns, join voting forces, etc.

9. As a group, students should discuss the questions on the "Round #2 Worksheet" (Handout 39). The note-taker is responsible for recording responses.

10. Following both rounds of the game, encourage a class discussion based on the worksheet questions and others that may arise. These are additional ideas for discussion:

      * The question of fairness may have come up. Was it fair that some persons' votes were worth more than others'? Did one person feel that others "ganged up" on him or her? What analogies of this experience to real life can be made?

* In real life, state officials decide where to put facilities like incinerators. How do citizens try to influence government? Does everyone have the same ability and resources to influence government? (Encourage students to consider what is needed to lobby government, organize groups, and draw publicity.) What resources of time and money are needed to do any of those actions? In what ways can a citizen influence an official's position?
* Are there some towns that want these facilities? Why or why not? (Help students understand that the promise of jobs, reduced energy costs, and other economic advantages are very important to some people.)
* What do you think is a fair way to decide where to site a hazardous waste facility?

Handout 36

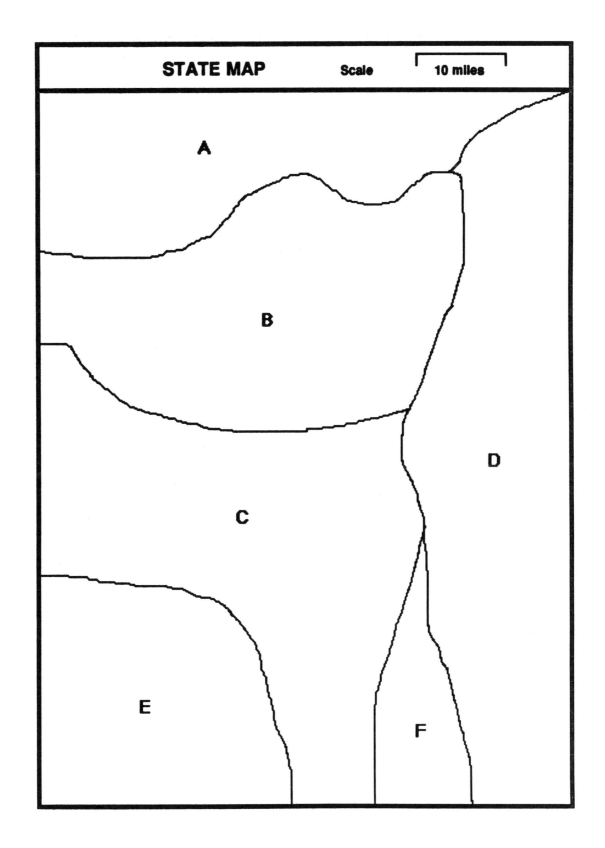

STATE MAP   Scale   10 miles

A

B

C

D

E

F

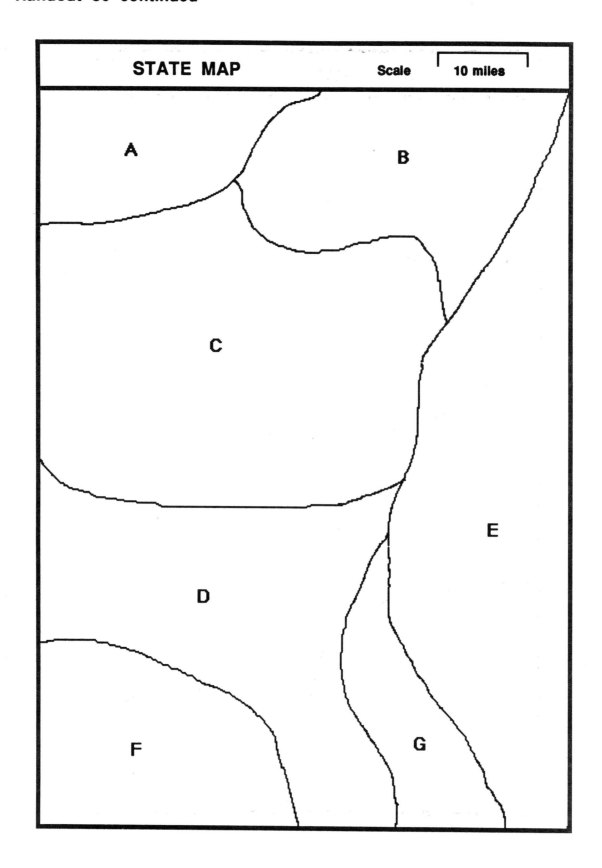

STATE MAP   Scale   10 miles

A

B

C

E

D

F

G

FACT SHEET

1. Within ten miles of a hazardous waste facility in another state, <u>slightly</u> higher rates of cancer have been reported.

2. The town that has the facility will have reduced electricity costs, estimated to be discounted 25%.

3. There will be a large amount of heavy vehicle traffic going to the facility.

4. One thousand new jobs will be available as a result of the facility.

5. Toxic ash from the incinerator will be deposited in a landfill adjacent to the facility.

6. The facility will be monitored constantly to meet EPA (Environmental Protection Agency) standards.

ROUND #1 WORKSHEET

Names of people in group:

1. Did your group come to a decision? If yes, what was the decision, and how did you arrive at it? If not, why were you unable to reach a decision?

2. What characteristics did any of the players give their towns? How did these factors relate to placing the incinerator?

3. How did you **feel** about this activity? Was it difficult? Why or why not?

ROUND #2 WORKSHEET

Names of people in group:

1. Describe what happened in your community after you were given the power cards. Were you able to reach a decision? If so, what was it, and how did you arrive at it? If not, why were you unable to reach a decision?

2. How did having different values attached to the votes affect the way the group discussed the issues?

3. How did you feel about this round and your "state"? Do you like your own and neighboring "towns"?

4. Do you think this activity has any relevance to the real world? Please explain.

TOWN CARDS

Copy and cut so that each group will have a set. Cut and distribute just enough for one card per student.

Town A	Town B	Town C	Town D
Town E	Town F	Town G	

POWER CARDS

Copy and cut so that each group will have a set. Cut and distribute just enough for one card per student.

10 Votes	1 Vote	3 Votes	9 Votes
7 Votes	5 Votes	2 Votes	

Activity 24

RESERVATIONS ABOUT WASTE

Objective

Through the performance of a skit, students will learn about a particular community's struggle to decide whether they should allow hazardous a waste incinerator to be built in their area. Students will also consider whether racism is a factor in the planning of such sitings.

Note

We have included two newspaper articles, "Navajos Vote Against Toxic Waste Facility," originally printed in the *Winslow Mail* in Winslow, Arizona, and "Indian Reservations Are Targeted for Nation's Waste," from the *Boston Globe*. These articles may provide useful background for you, and you may want to use it with your students after they have performed the skit about the tribe's decision regarding a harzardous waste incinerator on their land. The skit is based on information from HiTech Recycling, Inc. and from Greenpeace.

Suggested Previous Activities

11. What Should Be Done with All the Trash?
16. Household Hazardous Waste Audit
20. Hazardous Waste Incinerator Siting
22. Learning About Racism and Classism
23. In My Backyard?

Time

2 45-minute periods for skit writing
30-minute homework assignment (writing after skit)
1 30-minute class or homework assignment to read articles
20 minutes for discussion

Vocabulary

carcinogenic	racism
hazardous waste	reclamation
incinerator	recycling
landfill	

Materials

1 copy of Handout 40, "Reservations About Waste Skit: A Case from Dilkon Arizona," for each student
1 copy each of Handout 41, "Navajos Vote Against Toxic Waste Facility," and/or Handout 42, "Indian Reservations Are Targeted for Nation's Waste" for each student

Activity Procedure

1. Explain to students that they will be role-playing a town meeting taking place on a Native American reservation in Dilkon, Arizona. A private company, HiTech Recycling, Inc., is proposing the construction of a hazardous waste incinerator on a Native American reservation. The meeting is to decide whether or not to accept the proposal by HiTech Recycling, Inc.

2. The room should be set up so that the speakers, six tribal delegates, are in front of the class. Assign students to these roles. Also assign students to the roles of the chair of the meeting (Tom Top), the four guest speakers, and the medicine man. (There is a total of 12 speaking parts.)

3. All other students will act as tribal council members who will vote on the proposal.

4. Allow students a few minutes to review their roles and then begin the skit.

5. After the speakers have finished reading their parts, have the student who plays Tom Top lead an open discussion among all the tribal council members (i.e.,the rest of the class). After five to ten minutes of this, Tom Top should call for a vote and tally the results.

6. Students can work individually or in small groups to respond to the questions and tasks below. After students have had time to think about them, conduct a whole class discussion to share thoughts.

> * Write down three ways in which you agree with HiTech Recyling or feel their proposal will benefit the Navajo people.
> * Write down three ways that you think the facility will hurt the Navajo people.
> * What influenced your vote most?
> * Who spoke most convincingly? Why?
> * Do you think the planned siting of the incinerator involved racism? Do you think the Navajo are being discriminated against? Why or why not?

7. For homework, give students the assignment of writing newspaper articles that shows that the planned siting is based on racism and another article that shows that the decision is not based on racism.

Suggested Follow-up Activity

Have students read the article "Navajos Vote Against.Toxic Waste Facility" and/or "Indian reservations targeted for nation's waste". Discuss the article(s) in class using the following questions:

> * What new information did you learn from reading this article?
> * Does the article seem to favor one side or the other?
> * If you lived on a Native American reservation, how do you think you would respond to this controversy?
> * Do you think the Native Americans are being unfairly "dumped on"?

RESERVATIONS ABOUT WASTE SKIT: A CASE FROM DILKON, ARIZONA

Parts:

Chair of the meeting: Tribal Council head - Tom Top

Guest speakers:
    HiTech engineer - Janet Calcul
    Environmental spokesperson from HiTech - Sam Trim
    Native American subcontractor of HiTech - Bill Travator
    Greenpeace representative - Lisa Merced
    Tribal Council delegates with speaking roles:
    Tribal Council delegate - Kathy Honor
    Medicine man - Mike Mansie
    Tribal Council delegate - Sue Agny
    Tribal Council delegate - Suzy Quest
    Tribal Council delegate - Peter Real
    Tribal Council delegate - Barbara Gulliver
    Tribal Council delegate - Patrick Wright

Tom Top: We have called this meeting today to listen to a proposal from HiTech Recycling, Inc. for the construction of an incinerator. With us today from HiTech are Janet Calcul, an engineer, Sam Trim, an environmental specialist, and Bill Travator, a Native American employed by HiTech. Lisa Merced from Greenpeace is here to offer her perspective on hazardous waste incinerators. Sam Trim will be the first to speak.

Sam Trim: HiTech Recycling, Inc. has been in the business of building hazardous waste disposal sites for fifteen years. I drove two hours to come here today because I believe our proposal can provide economic opportunities for the Navajo people.

Peter Real *(stands and shouts)* : Give me a break! You don't care about the Navajo; you just want to make money!

Tom Top: Please sit down, Mr. Real, and let Mr. Trim speak.

Barbara Gulliver *(yells)*: Yeah, sit down! I want to hear what he has to say.

Sam Trim: Many people of the Navajo Nation are rightfully suspicious of the motives and reasons for this venture. No person of any race, creed or nationality would allow anyone to expose their families and children to any risk whatsoever. Those who oppose this venture make their case on emotional fears and ignorance of the facts. There is no zero risk in the world. I took a risk driving here today. But HiTech is committed to making its facilities as risk-free as is humanly possible.

Peter Real: Yeah, so why don't you be truthful and admit you are doing this only for the profit. If this thing is so safe, why don't you build it in your own backyard?

Tom Top: Mr. Real, if you cannot remain quiet, I will have to ask you to leave.

Sam Trim: HiTech is willing to provide two educational seminars for the Navajo Nation and to allow the young and old, men and women alike, to make an independent judgment of the facts. After this public educational and informative session, you decide as to whether HiTech shall continue or not. We are not trying to force anything on you. We believe this proposal will greatly benefit the Navajo people.

Suzy Quest: Why are you so interested in bringing this incinerator to our land?

Sam Trim: We see this as a great opportunity to help Native Americans. Bill Travator, a Native American with our company, can talk more about what HiTech has done for him.

Bill Travator: I am the chief executive officer of the Blaze Construction Company, the most successful Native American-owned company in the United States. I employ over 400 Navajo tradesmen on various reservations. As a Native American, I believe in the sacred care of the land from which we are all made. This proposed facility is safe and will provide hundreds of needed jobs. I believe the Navajo people deserve a chance for economic growth. Janet Calcul, our engineer, can tell you more about the safety of this project.

Janet Calcul: Let me start by saying that nothing is 100% safe. We take risks every day by driving our cars or just walking down a street. I can not tell you this recycling facility is 100% safe. This plant is a treatment, recycling, storage, and disposal facility. There will be .001% of residue left, but it will be sterile ash. We are talking about minute remnants. With the millions of dollars we are investing, it is in our interest to make a technically sound project.

Lisa Merced: Excuse me, Ms. Calcul, would you care to share how people's health has been effected by these "minute remnants"? How ground water has been contaminated? How unusually high rates of cancer have been reported in people living near hazardous waste incinerators?

Janet Calcul: As I said, Ms. Merced, nothing is risk-free. Many of these statistics you quote are the result of problems with early designs of incinerators. Technology has greatly improved in this area. As Mr. Trim said, we have been in this business for fifteen years, and we have been very successful.

Lisa Merced: What is your measure of success? The number of people who do not get cancer? Are you willing to live next to one of your facilities?

Tom Top: I would like to let HiTech finish their presentation. Ms. Merced, we will hear from you shortly. Was there a question from a delegate?

Suzy Quest: Yes, what can the Navajo people expect to gain from this industrial development project?

Sam Trim: This plan for development can provide needed employment, training, education, higher living standards, improved schools and social programs, scholarship opportunities, improved medical facilities, and, most importantly, a positive, progressive program of Navajo administrative leadership. Don't believe the scary rumors. Unskilled and unqualified terrorists spread horror stories of improperly managed facilities.

Lisa Merced *(stands up and shouts)*: Terrorists? You call people who are educating others about the risks of toxic incinerators terrorists?!?!! There is ...

Tom Top: Please, Ms. Merced. I would like everyone to get a chance to present their view. Please be patient. *(Lisa Merced sits down.)*

Sam Trim: Our personnel go through extensive training programs, and in fifteen years we have never had a problem. All peoples, races, and nations on earth are responsible to the next generation for its survival on this planet. We can ill-afford not to take advantage of the financial opportunities as well as the credit for saving the land from continued environmental damage. We must be committed to reclamation and to recycling wastes. Our own survival should be reason enough without the financial and socio-economic opportunities that accompany this industrial development venture.

Sue Agny *(jumps up and says in an angry voice)*: Now wait a minute, you give us all this information, but why do the Navajo need to be involved in white people's problems of environmental reclamation?

Sam Trim: We all share in the responsibility and care of our earth. The first and most important reason is that Navajo and Anglo all live in a world that is growing smaller every day. The quality of our environment is changing for all races and nations. Any Navajo that owns a car or truck contributes to the environmental pollution. Every Navajo that uses a plastic bag to carry his or her groceries home has caused one teaspoon of hazardous materials to be made.

Patrick Wright *(stands up)*: So what? Just because we use products that contribute to environmental pollution doesn't mean we should take the burden of all the white people's waste as well. You white folks are always trying to dump on us, and I don't trust any of you.

Barbara Gulliver *(stands up)*: Mr. Trim has a good point. As Native Americans, we have always been caretakers of the earth. But technology and a growing society have put tremendous stress on the environment. Here is our opportunity to help out. I think we are being offered a good deal, and we should not pass it up.

Sam Trim: I think this is a wise choice. I personally understand why some people feel cautious and fearful in supporting this proposal. I myself faced a similar choice several years prior to pursuing a career in this field. I was extremely fearful of my first exposure to hazardous circumstances and materials. This fear was caused by ignorance and misconceptions. I didn't believe that I personally could make a difference in the problems and ills of this world. As I gained in wisdom and knowledge in this unexplored new field, I realized I had been wrong. I have made a difference.

Lisa Merced: May I speak now?

Tom Top: Yes, Ms. Merced.

Lisa Merced: When I was first told of this facility, it was presented as a recycling center, but after some persistent questioning, I found it was a Class 1 toxic waste landfill. You should know some facts about these facilities and about why Native American land is so often selected.

Native American communities are prime targets for the following reasons: First, the communities need jobs. The companies are offering them, but the number of jobs they claim to offer qualified applicants is often exaggerated. Second, the companies have offered token amounts of money to tribal councils if they accept an incinerator or landfill. Third, waste companies are billing their facilities as state-of-the-art facilities that are harmless to the environment. But we know this is not true! Fourth, Native American nations are independent nations. This means they do not have to follow state environmental regulations. The waste companies also offer assurances that they will follow the federal standards set by the Environmental Protection Agency, which leads people to believe that the federal government wouldn't allow the companies to exist if they poisoned the environment. Don't be fooled by Mr. Trim's claim that they are thinking of the Navajo people's economic advantages. They are able to beat environmental regulations by using reservation land.

Kathy Honor (*stands up*): Ms. Merced, have you ever gone to bed hungry? Do you wonder where and when you will get your next meal? I don't think you understand our economic hardships. Our unemployment rate is 70%. Those of us who do have jobs have to travel far outside our community. We've got to do things on our own land. We can't afford for all our college students to just disappear to the bigger cities. I don't like to see my people depending on others all their lives. HiTech is offering us many economic advantages. How can we pass this up? Look, it says here that they're offering an annual trust fund of $100,000 to $300,000 for our community to use at our own discretion, a training program for our residents about health, safety, and emergency services in case of an accident, and training for employee advancement. They estimate that it would generate an annual payroll of $3 million.

Lisa Merced: You are being lured by their promises. Why do you think they need to offer so many economic enticements? Why are they willing to do so much? I'll tell you why. Because state laws don't apply on reservations, and state laws are stricter than federal laws. A recent study done by a private company in California showed that low income and minority areas are the best place for incinerators because people in those groups are less likely to protest. You are being taken advantage of.

They say their facility will be safe. Safe for whom? They don't live here. In the short term you may get jobs, but is it worth it in the long run when we know that other communities have suffered permanent damage? Scientists agree that hazardous waste landfills can contaminate ground water and put chemicals into the air. They also agree that there are short term and long term health effects from accumulations of toxic fallout from incinerators. In an EPA report, scientists warned that the human health risk from incineration of carcinogenic heavy metals such as chromium, cadmium, and beryllium may be significant. Incinerators release unknown quantities of unknown chemicals and present health threats of unknown magnitude to the people in surrounding communities.

Mike Mansie: The Navajo Nation has a long tradition of living in harmony with the land. We respect our land. As part of our history, we use fire to heat ourselves and cook our food. We have ceremonies where we see and honor fire as a friend. We must keep fire clean. But this technology does not do that. This is not magic. The residue goes up in the smoke and it's still poison, or it sits in the ground and contaminates our water.

Tom Top *(speaking to all tribal delegates and members assembled):* You have heard from many people and now we must discuss amongst ourselves what should be. Do any of the delegates who have not yet spoken wish to express their point of view?

(Tom Top leads a general discussion.)

Later:

Tom Top: Our time is running out, and it is time to take a vote. Please consider both sides as you think about how to vote.

(The delegates, silent and speaking roles alike, now decide how to vote.)

Tom Top: First, those in favor of the facility raise your hands. (*He counts silently.*) Now, those opposed raise your hands. (*He counts silently.*)

(Tom Top announces the totals and states whether the facility plans will proceed or not.)

Navajos Vote Against Toxic Waste Facility

FROM: "Winslow Mail"
May 4, 1990.
Winslow, Arizona

By Walt Olhava

Greg Shepard, a Bureau of Indian Affairs employee and treasurer of Citizens Against Ruining Our Environment (CARE), said what was proposed was not, in fact, what was going to be built.

"In December of 1988, we heard of plans for a waste dump," he said. "It was presented at a chapter meeting as a trash dump, and it was approved."

He said that the tribe was offered enticements such as an annual trust fund of $100,000 for the community to use, a comprehensive training program to deal with accidents and training for employees to advance their skills.

"When the residents heard that, they said 'Yes, now we can have our children educated with that kind of money. Lets do it'

"But we researched it and found out that it really was a toxic waste dump," Shepard said. "Then, in February of 1989, we voted against it. We thought it was a dead issue."

Bill Aubrey, president of Hi-Tech Recycling, called the plant a treatment, recycling, storage and disposal facility and said it was harmless. Gary Stevenson, a spokesman for Waste-Tech Ser-

vices, a Colorado company that provides HiTech Recycling with the engineering and design for waste disposal sites, said they were talking in terms of .001 percent of residue left in the form of sterile ash and minute remnants.

"So there was a division of two sides, those who were in favor of the promise of jobs and income and those who were on the side of tradition," Shepard said. "The Navajo culture has strict laws that say that having something like this is totally against the peoples' laws and completely unwanted. So it was voted against."

George Joe, a member of CARE, agreed with that assessment.

"The people voted that way because the project was not in harmony with the earth," he said. "Dangers existed in it and the people realized those dangers."

The environmental group Greenpeace filmed and aired a video presentation on the matter. In it, Steve Maupin, a representative of one of the toxic waste companies, called the project one for "salvation of the waste, a rejuvenation of the waste into a reusable product," at a meeting on the matter.

Then the video showed Bradley Angel, a Greenpeace representative, saying that he was at first told it was a recycling center, but, after 20 minutes of persistent questioning, finally found out it would be a Class 1 toxic waste landfill.

That was followed by a collection of Dilkon residents, Arthur Yazzie being one of them, saying that all the people who live nearby were opposed to the project.

Work to combat the placing of toxic landfills on reservation land is being done at various levels. Paul Rodarte, director of Citizen Alert Native American Program in Reno, Nev., is a Paiute and Shoshone Native American from the reservation in Stillwater, Nev.

He is currently in Washington, D.C. lobbying to put an end to locating toxic waste dumps on reservations. He said the issue is difficult to fight for several reasons.

"Indian communities are prime targets for the following reasons,"

Dilkon Issue

Continued From Page 1

he said from the nation's capitol.

"1. The communities need job and the companies are offering them, although the number of job offered to qualified applicants is often inflated.

"2. The companies have offered token amounts of money to tribal councils if they accept an incinerator or landfill.

"3. Waste companies are billing their facilities as state-of-the-art facilities and harmless to the environment.

"4. Indian nations are sovereign nations exempt from the jurisdictions of state environmental agencies. The waste companies also offer assurances that they will follow standards set by the EPA, which leads people to believe that the federal government wouldn't allow the companies to exist if they poisoned the environment."

But the Navajo Nation has a long tradition of living in harmony with the land and Rodarte said one of the ways the tribe sought an answer to the question was by looking at the history of fire in the Navajo tradition.

"We really respect fire," he said. "As part of our history, we use fire to heat ourselves and our food. We have ceremonies where we see and honor fire as a friend. We must keep fire clean. But this technology does not do that. This is not magic. The residue goes up in the smoke and it's still poison."

Rodarte said the way this will be solved is through working at source reduction and "the people working together to take care of Mother Earth."

Indian reservations are targeted for nation's waste

By Larry Tye
GLOBE STAFF

From Connecticut to California, disposal companies have come up with a novel scheme for handling wastes so rancid or toxic that no city or town will take them.

Give them to the Indians.

The firms have pursued deals with the Sioux, Navajo, Choctaw and more than a dozen other tribes – for understandable reasons. Reservations

offer 53 million acres of open space along with exemptions from state environmental laws and most state and local taxes. And gaining preliminary approval can be as easy as winning over tribal elders so anxious to ease joblessness that many beg the companies to come.

Even the federal government is drawn to the reservations as a possible storage site for the nation's longest-lasting radioactive garbage.

But the dump proposals are raising increasingly troubling questions for American Indians

and for Americans generally. Indians risk scarring their cherished landscape with dumps that often leak and seldom deliver promised economic rewards, while the rest of the country in effect shunts the dumps it rejects onto those less able to resist.

"They've got to find a place to dump it and probably the easiest is to dump it on the Indian people. We don't have the resources to deal with it or the political power," said William Koenen, a

INDIANS, Page 12

Dump sites proposed for Indian reservations

NOTE: Barrels inside states do not represent exact location of reservations.

ALASKA
- Chickaloon Village rejected hazardous waste incinerator.
- Tyonek tribe approached for hazardous waste incinerator.
- Coshutta tribe approached for hazardous waste facility.

ARIZONA
- Navajo community of Dilkon rejected hazardous waste incinerator.
- Tohono O'Odham Nation approached for garbage incinerators.
- Pasqua Yaqui tribe, Gila River Indian Reservation, and Salt River Pima-Maricopa Indian Community rejected facility to process solid waste.
- Pima-Maricopa rejected medical waste incinerator.
- Paiute-Kabab tribe rejected hazardous waste incinerator.
- Fort Yuma Reservation rejected landfill proposals.

CALIFORNIA
- Jackson Rancheria rejected medical waste incinerator.
- Morongo Indian Reservation rejected garbage dump.
- Los Coyotes Reservation rejected solid waste dump.
- Campo Band Reservation approached for a solid waste landfill. Tribal council negotiating.
- La Posta Reservation approached for hazardous waste incinerator and recycling facility. Permit process underway.
- San Isabel Reservation approached for waste disposal facility.
- Borona Reservation rejected disposal facility plan.
- Hopland Rancheria approached for tire cogeneration plant.
- Picayune Reservation approached for cogeneration plant.

CALIFORNIA, CONT.
- Cortina Reservation considering asbestos disposal facility.

COLORADO
- Colorado Indian tribe approached for carbon recycling facility.
- Pine Ridge Reservation received at least 10 proposals for solid and hazardous waste disposal facilities.
- Ogala Sioux tribe on Pine Ridge Reservation approached for sewage sludge ash and asbestos disposal plant.
- Pine Ridge Agency offered hazardous, solid, and nuclear waste disposal and nuclear power plants.

FLORIDA
- Seminole Indians approached for hazardous waste incinerator.

MISSISSIPPI
- Choctaw tribe rejected hazardous waste dump.

MONTANA
- Northern Cheyenne Reservation approached for waste conversion and recycling program.

NEVADA
- Moapa band of Southern Paiute Indians approached for hazardous waste incinerator.
- Washoe tribe approached for garbage incinerator and/or landfill.

NEW YORK
- Mohawks received numerous proposals for hazardous waste, sewage sludge and medical waste incinerators, and landfills.

NORTH CAROLINA
- Cherokee tribe rejected incinerator.

NORTH DAKOTA
- Fort Berthold Reservation considering incinerator.

OKLAHOMA
- Seminole Nation rejected medical waste incinerator.
- Kaw tribe rejected hazardous waste incinerator.
- Pawnee tribe considering contract for facility to compost sludge.
- Tonkawa tribe approached for railroad off-loading facility for hazardous waste.
- Cherokee tribe rejected waste disposal facility.
- Chickasaw tribe approached for nuclear waste dump.
- Chilloco Development Authority of the Kaw, Ponca, Pawnee, Tonkawa and Otoe tribes approached for waste disposal plants.

RHODE ISLAND
- Narragansett tribe turned down incinerator.

SOUTH DAKOTA
- All South Dakota reservations offered plan using cement kiln dust and local clay soils to make adobe building blocks.
- Pine Ridge Reservation of South Dakota approached for solid waste incinerator.
- Rosebud Reservation considering solid waste, incinerator ash and sewage ash landfill.

UTAH
- Goshute and Ute tribes considered but did not pursue proposals for hazardous waste incinerators.

WASHINGTON STATE
- Tulalip tribe considering municipal solid waste incinerator.
- Puyallup tribe considering waste disposal facility.

Chippewa and part of a growing Indian movement whose campaign against disposal companies so far has prevented any dumps from opening.

Sen. Thomas Daschle, Democrat of South Dakota, is equally disturbed: "The United States once again is failing to live up to its responsibilities. We look for the easiest solution, and often that means somebody else pays."

Yet Raymond Oneglia, president of O&G Industries of Torrington, Conn., sees the issue differently.

"Is there anyplace around Boston within 100 miles that would accept a landfill," asked Oneglia, who wants to build a 5,700-acre landfill on South Dakota's Rosebud Sioux reservation, a million acres of rolling plains, meandering streams and roaming antelope.

"The Indians want it badly ... they called us," said Oneglia, whose plan has become a lightning rod for antidumping activists. "You've got to go out there and see how those Indians live. ... You cannot believe how poor they are."

Oneglia is not the only one interested in bringing jobs, money and garbage to the 1 million Indians on more than 500 reservations nationwide.

Since 1988, according to some groups, tribes have been given as many as 120 plans for waste dumps. "I doubt there's any reservation in the country that has not at least been approached," said Whit Fosburgh, an aide to Daschle.

Sometimes, the interest has come from engineering giants such as Bechtel Enterprises, whose lawyers wrote the Bureau of Indian Affairs last year expressing interest in putting landfills and energy plants on Indian land.

According to George Farris, environmental chief at the bureau, "fly-by-night companies that can't afford to do these sorts of things" also have submitted plans for dumps handling everything from trash to toxins.

At first, most tribal leaders were very receptive. After all, unemployment on reservations generally ranges between 60 and 80 percent. And because tribal members often lack technical training, land is the main carrot that the leaders can offer developers.

In case after case, however, as word of the negotiations leaked out, opposition snowballed.

Safety was one sore point. State environmental agencies and local health departments typically review landfill proposals and police the dumps, but neither group has authority over reservations, which in many ways operate as sovereign nations.

Dumps on Indian land must be approved by the US Environmental Protection Agency and the Bureau of Indian Affairs, a process that can cost $3 million and take three years. But the National Congress of American Indians complains that those agencies cannot be trusted since they have failed to clean up crumbling asbestos, leaking oil tanks and other pollution already present on Indian land.

"If EPA does with the reservations what it did with the Superfund toxic cleanup program, we're in big trouble," Daschle said, alluding to repeated delays in cleaning hundreds of toxic dumps.

The Indians' biggest gripe with the proposed landfills is the way they would transform the land itself – which to Indians is a spiritual birthright and the place where, after they are buried, new life takes root.

"We love this land. Our message to the corporations and government entities behind these dumps is that the Lacota people are going to stand up and fight," said JoAnn Tall, who helped defeat a waste facility proposed for her Pine Ridge Reservation in South Dakota.

Concerns such as those, passionately voiced by activists, have defeated dozens of proposed dumps from Arizona to Utah and New York to Rhode Island. Wanona Laduke, a Chippewa from Minnesota, helped assemble more than 500 Indian organizers this month in the Black Hills of South Dakota to celebrate those successes.

But Laduke and the others made clear the threat is

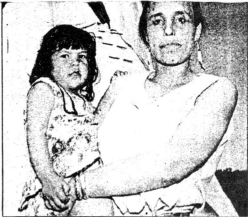

AP PHOTO

Wanona Laduke, a Chippewa activist, holds her daughter, Waseyabin.

not over, as plans for one toxic and two trash landfills are moving ahead in California, and the Department of Energy has written to 565 tribes asking them to accept radioactive wastes. The fiercest battle of all, however, is being fought over O&G Industries' proposal for Rosebud.

The Rosebud scheme is drawing attention partly because it would be one of the nation's biggest landfills. But it also embodies most of the other concerns voiced by Indians nationwide.

First, there is the contract the tribal council signed with an O&G subsidiary. While it promises per-ton payments that could reach millions of dollars a year, along with jobs, it adds that "in no event shall any environmental regulations or standards of the state of South Dakota be applicable to this project."

It also says the firm will conduct all the monitoring of groundwater under the landfill. And if the tribe adopts new environmental standards or raises taxes, the disposal company must be exempted, or the cost of compliance paid from the tribe's profits.

What kinds of wastes would the dump accept? That, too, is up to the disposal company, although toxic, nuclear and medical wastes are banned. "We don't know exactly what would be brought in by rail. Whatever business accumulated, it's a business venture," said Oneglia, O&G's president. "We'd take it from Boston if we could," he added.

Many Rosebud Indians also are outraged that an old cemetery would be surrounded on three sides by garbage. They worry that O&G, a multimillion-dollar firm that operates concrete plants and quarries, lacks experience in building and running landfills. They accuse the company and tribal leaders of forcing the project down their throats with high-powered lobbyists such as James Abourezk, the former US senator from South Dakota.

Those concerns could play a decisive role in tribal elections set for August, and those elections could determine the fate of the O&G plan. If the current council is reelected, the Bureau of Indian Affairs and the EPA would review the proposal. If it is not, dissidents – whose rallying cry is "a vote for an incumbent is a vote for out-of-state trash" – vow to kill the dump.

Yet while technical issues are important, Rosebud activists, like those elsewhere, say it is the violation of their land that makes them angriest and should ensure their victory at the polls.

"There was a white guy stumbling around in the Atlantic Ocean and when he happened on this country we had 100 percent of this land. Now we have 3 percent. Why do you have to use that for your trash," asked Ron Valandra, a leading dump opponent.

"To people back East, this land would look desolate and unproductive," he added. "But it's our land, and we don't think we should use it to take care of a society that doesn't care how much trash it generates."

Activity 25

UNCOVERING ENVIRONMENTAL INJUSTICE

Objective

Students will learn about the relationship between racism, classism, and environmental decision-making in this country by examining the demographics and experiences of communities chosen as sites for toxic waste disposal.

Suggested Previous Activities

21. Where Does Hazardous Waste from Your State Go?
22. Learning About Racism and Classism
23. In My Backyard?
24. Reservations About Waste

Time

6 class sessions (or 4 class session and 2 homework assignments)

Vocabulary

aquifer	landfill
classism	leachate
demographics	PCB's
environmental racism	racism
incinerator	toxic

Materials (copies for each group or each student if for homework)

Handout 43, "Landfill Area Demographics"
Handout 44, "Toxic Wastes and Race in the United States"
Handout 45, "Questions About Landfill Area Demographics"
Handout 46, "Residents' Attitudes"
Handout 47, "Questions About Residents' Attitudes"
Handout 48, "Two Southeastern Cases Studies"
Handout 49, "Case Study Questions"
Handout 50, "Dumping on the Poor"

Activity Procedures

Introduce this activity with the following statements and question, keeping them in mind throughout the activity.

> * Toxic waste is a problem now and will probably continue to be a problem in the future, even with source reduction. Toxic waste must go somewhere. What factors should be taken into account in deciding where it should be disposed of?

1. Divide students into groups of three or four. Give them Handouts 43, 44 and 45. In small groups, students will analyze the charts and map, and answer the questions. As a whole class, discuss the groups' answers.

In discussing questions students may realize that regional statistics or state-by-state statistics might offer more detailed comparisons. They might ask about the size of each area and how close residents' homes were to the landfill sites. They might be interested in the 1990 data. Or they might be curious about the percentage of registered voters in the population compared with state, regional, or national percentages. There are many possible questions, any of which could be excellent topics for further research. The Census Bureau has data on these topics (see resource directory).

2. Have students read and discuss in small groups Handouts 46 and 47 on Residents' attitudes. Discuss questions and responses with the whole class.

3. In class or for homework have students read Handout 48, "Two Southeastern Case Studies," and think about the questions (except the last two) on Handout 49. In small groups, have students answer the questions. Discuss group answers as a whole class.

4. Give students Handout 50, the article,"Dumping on the Poor." After students have read the article, have them answer the last two questions on Handout 49 in small groups, and then discuss their answers as a whole class.

Suggested Follow -up Activities

Is there environmental racism in your community? Have students use the information from Activity 21, "Where Does Hazardous Waste From Your State Go?," and have them find out who lives near the disposal sites. Is it near a town line? Why do students think that particular location was selected?

For further study, you might wish to contact Waste Management, Inc., the Commission on Racial Justice, and the National Toxics Campaign to find out if there are any recent developments in either of the two disposal case studies, Emelle and Warren County.

Handout 43

LANDFILL AREA DEMOGRAPHICS

Below you will find two charts from the United States General Accounting Office, taken from *Dumping in Dixie: Race, Class, and Environmental Quality,* by Robert D. Bullard. They are the result of a study done in 1983 to examine the placement of landfills and hazardous waste facilities in the United States.

Landfill (State)	Population Number	Population % Black	Median family income ($) All Races	Median family income ($) Blacks	Population below poverty level Number	Population below poverty level %	Population below poverty level % Black
Chemical Waste Management (Alabama)	626	90	11,198	10,752	265	42	100
SCA Services (South Carolina)	849	38	16,371	6,781	260	31	100
Industrial Chemical Co. (South Carolina)	728	52	18,996	12,941	188	26	92
Warren County PCB Landfill (North Carolina)	804	66	10,367	9,285	256	32	90

*Areas represent subdivisions of political jurisdictions designated by the census for data gathering.

Source: U.S. General Accounting Office, *Siting of Hazardous Waste Landfills and Their Correlation with Racial and Economic Status of Surrounding Communities* (Washington, D.C.; General Accounting Office, 1983), p.4.

Operating Hazardous-Waste Landfills in the Southern United States and Ethnicity of Communities (1987)

Facility Name	Current Capacity in Acre-Feet*	Percent of Population in Zip code Area Minority	Percent of Population in Zip code Area Black	Percent of Population in Zip code Area Hispanic
Chemical Waste Management (Emelle, AL)	30,000	79.5	78.9	0.0
CESCO International, Inc. (Livingston, LA)	22,400	23.8	21.6	1.8
Rollins Environmental Services (Scotlandville, LA)	14,440	94.7	93.0	1.5
Chemical Waste Management (Carlyss, LA)	5,656	6.8	4.6	1.7
Texas Ecologists, Inc. (Robstown, TX)	3,150	78.2	1.6	76.6
GSX Services of South Carolina (Pinewood, SC)	289	71.6	70.5	1.1
US Pollution Control, Inc. (Waynoka, OK)	118	37.3	23.2	12.3
Gulf Coast Waste Disposal Authority (Galveston, TX)	110	4.3	0.0	3.3
Rollins Environmental Services, Inc. (Deer Park, TX)	103	7.3	0.3	6.2

Reprinted by permission of Westview Press.

193

TOXIC WASTES AND RACE
In The United States

A National Report on the Racial and Socio-Economic Characteristics of Communities with Hazardous Waste Sites

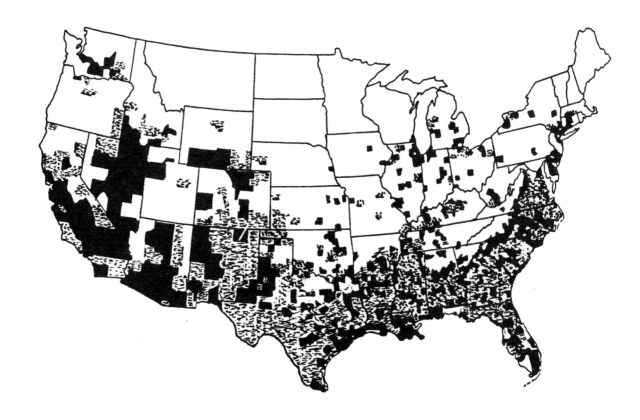

KEY:

Shaded areas represent counties where the Black and/or Hispanic percentage of the population is greater than their respective national percentages.

(Black percentage of the population in U.S.: 12 percent)

(Hispanic percentage of the population in U.S.: 6 percent)

Dark areas represent counties where the Black and/or Hispanic percentage of the population is greater than their respective national percentages and where five or more uncontrolled toxic waste sites are located.

Source: U.S. Bureau of the Census

U.S. Environmental Protection Agency

Reprinted by permission of the Commission For Racial Justice, United Church of Christ

QUESTIONS ABOUT LANDFILL AREA
DEMOGRAPHICS

Names of people in group:

Appoint a note taker in your group. Discuss the questions below, referring to the charts on the "Landfill Area Demographics" sheet and to the map entitled "Toxic Wastes and Race in the United States." It is important to note that, according to 1980 census data, African American people were 12% of the total U.S. population and Hispanic people were 6% of the population.

1. What patterns do you notice on the map?

2. How many of the sites on each chart are located in areas where the percentage of Blacks or Hispanics in the population is <u>higher</u> than the national percentage?

3. How many of the sites on each chart are located in areas where the percentage of Blacks or Hispanics in the population is <u>lower</u> than the national percentage?

4. The median family income in 1980 for all families was $19,917, while for Black families the median family income was $12,598.

a. How many sites on the first chart are located in areas where the median family income is <u>lower</u> than the national average for all races?

b. How many sites on the first chart are located in areas where the median family income is <u>lower</u> than the national average for Blacks?

5. How many sites on the first chart are located in areas where the median family income is <u>higher</u> than the national average for all races? How many sites on the first chart are located in areas where the median family income is <u>higher</u> than the national average for Blacks?

6. The national percentage of the population living below the poverty level in 1980 was 12.4% for all races and 29.9% for Blacks. How do the sites on the first chart compare to the national statistics?

7. What conclusions might you draw from this information?

8. If you represented a landfill company wanting to establish a new site in an area similar to those on the charts, what arguments would you use to persuade local residents to accept the landfill?

9. What other questions do you have or information do you need in order to draw conclusions about the placement of both landfills and hazardous waste facilities in the United States?

RESIDENTS' ATTITUDES

Below is information from *Dumping in Dixie: Race, Class, and Environmental Quality,* by Robert D. Bullard. Mr. Bullard interviewed 523 residents who were randomly selected from African American household heads.

TABLE   Distribution of Responses to Items Assessing Economic Versus Environmental Trade-Offs (percent)

Items	Emelle (N = 100)
Individuals in this community accept health risks as a trade-off for jobs	
Agree	49.5
Disagree	50.5
We should think of jobs first and environment second	
Agree	41.0
Disagree	59.0
Employment opportunities for local residents have improved with the facility	
Agree	49.5
Disagree	50.5
The facility has generated needed tax dollars for the community	
Agree	61.6
Disagree	38.4
The benefits that the community derive from the facility far outweigh the negatives	
Agree	38.2
Disagree	61.8

Handout 46 continued

Table   Reasons Respondents Feel Their Community Was Singled Out in the Facility Siting Process (percent)

Reasons[a]	Emelle (N = 47)
Race of residents	29.8
Residents are poor and need jobs	23.4
Residents are powerless	8.5
Residents are unorganized	10.6
Land values and location	27.5

[a]The above categories represent the responses to an open-ended item asked only of those individuals who believed their community had been singled out for the facility.

Table   Residents' Attitudes on Facility Siting Equity (percent)

Attitudinal Items	Emelle (N = 100)
Feel it was fair to locate facility in community	
Yes	28.0
No	72.0
Feel community was singled out to receive facility	
Yes	51.0
No	49.0
Feel angry to have facility in community	
Yes	57.0
No	43.0
Have become less angry over time	
Yes	22.6
No	77.4
Have come to accept the idea that the facility will likely be in the community for some time	
Yes	88.0
No	12.0

QUESTIONS ABOUT RESIDENTS' ATTITUDES

Names of people in the group:

Appoint a note taker from your group and answer the questions below.

1. Write sentences that describe each of the most widely held opinions.

2. Write sentences that describe each of the opinions about which people surveyed were sharply divided.

3. If you represented the company that owned the landfill in Emelle, would you be concerned about residents' opinions? If so, what arguments would you use and what actions would you take?

4. If you were running for County Commissioner in Emelle, what would your platform be regarding the landfill?

5. If you could meet the people from Emelle who answered these questions, what would you want to say to them? What questions would you have?

TWO SOUTHEASTERN CASE STUDIES

Below are short case studies about two hazardous waste disposal sites. These sites are landfills in Warren County, North Carolina and Emelle, Alabama. These particular landfills are important because the disputes related to the choice of these sites helped make the public aware that many of our toxic waste sites are found in poor and minority communities. Because the materials in these disposal sites are hazardous, people in the communities surrounding the sites often feared they would have greater air and water pollution problems that would threaten the health of people living in the communities.

Locations for hazardous waste disposal are selected by state governments or by disposal companies with the approval of state governments. People are becoming more and more aware that toxic waste disposal sites in their communities might threaten their health, and people often fight to keep disposal sites out of their neighborhoods. These protests make it more and more difficult for state governments to find locations to dispose of toxic waste. Many times promises of employment opportunities at the disposal facility tempt poor communities to support a site because they hope to bring jobs to their people. Often, however, such jobs never materialize.

Since the public has become more aware of the facts about how toxic waste disposal is affecting minority and poor communities, many groups have sprung up within communities to help organize against sitings for toxic disposal in their areas. Sometimes people from communities experiencing problems created by toxic waste try to help other communities learn and organize to protect themselves. One such national organization, Citizens Clearinghouse for Hazardous Waste, was begun by Lois Gibb, a resident of Love Canal, New York, where uncontrolled toxic waste dumping threatened the health of her community.

Case Study 1: Warren County, North Carolina

In 1982, a location was selected in Warren County, North Carolina as a disposal area for PCB's. PCB's are polychlorinated biphenyls, a substance formerly used in electrical transformers, insulation for electric cables, and the production of electric condensers. PCB's were banned in 1977 when it was discovered that they build up in the food chain and pose health threats. In 1978, PCB's were illegally dumped along the sides of roads in North Carolina, contaminating 241 miles of roadside earth that had to be dug up and disposed of. The state considered 90 sites and finally selected one in Warren County, North Carolina.

The population of Warren County was 66% African American, whereas African Americans comprised only 12% of the whole United States population.

Ninety percent of the African American people living in Warren County live below the poverty level. This demographic information provides some of the evidence that protesters believe shows that toxic disposal sites are disproportionately placed in poor and minority communities.

The Environmental Protection Agency (EPA) used money provided by the "Superfund" to clean up and dispose of the PCB-contaminated soil in North Carolina. The Superfund is a special fund used to clean up the most dangerous hazardous waste sites in the country. Once the state of North Carolina had chosen the Warren County site, the EPA studied it. They said the site was an acceptable place for the disposal of the soil because it met geological requirements made by the Toxic Substances Control Act (TSCA) and other requirements made by the state. The TSCA regulations for PCB's require that "a landfill be in an area of low to moderate relief [relatively flat], with silt and clay soils, and above the historical groundwater table. The State wanted the landfill to be in an area (1) bounded by the counties where the PCB spills had occurred, (2) having a minimum of 16 acres, (3) isolated from highly populated areas, and (4) accessible by road with a deeded right-of-way" (GAO, 1983, p. 9).

The building of the facility was held up by two lawsuits. One lawsuit was brought by the Warren County commissioners and the other by the local chapter of the National Association for the Advancement of Colored People (NAACP) and the United Church of Christ Commission for Racial Justice. The first lawsuit was settled by an agreement made between the state and the commissioners.

It was the second lawsuit, claiming that racial discrimination was a motivating factor in the choice of the site, that brought the Warren County disposal site to national attention. A series of marches, demonstrations, and a non-violent civil disobedience campaign resulted in over 500 arrests. In the end, the judge stated that, "There is not one shred of evidence that race has at any time been a motivating factor for any decision taken by any official," and with that finding, construction of the landfill moved ahead.

While the protesters did not succeed in preventing the landfill from being built, their movement received national attention. U.S. Representative Walter Fauntroy participated in the demonstrations and was one of those arrested. After the campaign, Rep. Fauntroy requested that the United States government's General Accounting Office (GAO) study the communities in the southeastern United States where hazardous waste disposal sites existed. This report showed that the people who lived in all of the communities where sites existed were predominantly poor and African American.

Later in 1987, the United Church of Christ released a study of all communities surrounding toxic waste disposal sites. This study reinforced the findings of the GAO report that such sites are disproportionately found in poor and black or Hispanic or Native American communities (see map "Toxic Wastes and Race in the United States").

The people who participated in the protests against the siting of the Warren County disposal facility were disappointed that they were unable to prevent the landfill from being built, but they believe that the national attention generated by their campaign helped make others aware of the existence of what the Commission on Racial Justice has called "environmental racism." In addition, some of the participants in the protest believe that without their campaign, the Warren County landfill might have grown into a larger PCB disposal site, taking in PCB's from other parts of the country. As it is, the landfill was closed once the contaminated soil from North Carolina roads was deposited there.

Case Study 2: Emelle, Alabama

Emelle is located in Sumter County in western Alabama. African Americans make up 69% of the county's population. More than 33% of the population live below the poverty line (*Dumping in Dixie* Bullard, 1990, p. 69). Historically, the area was largely an agricultural community, but farming began to decline in the 1970's. "The demise of agriculture has left a 20% countywide unemployment rate and a significantly higher black unemployment rate. It is not uncommon in many black communities in the county to have one-third of the workforce unemployed. These bleak economic conditions and cheap land made the county a likely candidate for polluting industries -- especially waste disposal companies" (Bullard, 1990, p. 70).

In 1977, a group of local businessmen, including the son-in-law of then governor George Wallace, acquired 2,400 acres of land and received a permit to establish a toxic waste disposal facility. Some of the concerned citizens of the area question whether proper procedure was used in obtaining the permit and allege that the former governor's son-in-law used his political connections to obtain the permit. In 1978, the facility was purchased by Waste Management, Inc., the largest waste disposal company in the United States. Waste Management purchased additional land, bringing the total to 3,200 acres and making it the largest toxic waste disposal site in the world.

The location was selected because it has unusual geological qualities. "Much of west Alabama is underlined by a layer of limestone called Selma Chalk which averages 700 feet in thickness in Emelle." The chalk is highly impermeable. According to Booth Grunter and Mike Williams, two environmental reporters, the company claims that "it would take 10,000 years for waste escaping from the landfill's trenches to penetrate to the aquifer below the site" (Bullard, 1990. p 70).

In 1984, strict federal guidelines were put into effect, and the new parts of the landfill had to be lined with plastic. Nevertheless, although the owners believed that the chalk layer would never leak, their own test wells showed contamination in 1985.

Highway 17, a farm-to-market road, leads to the facility. Large disposal trucks go back and forth 24 hours a day past the homes of the area residents. In

1986, the toxic wastes from 45 states and 6 foreign countries ended up in Emelle. Emelle receives some of the most hazardous materials. In 1989, it received 700,000 tons of waste.

Some people feel the facility was brought into the community without any input from the residents. According to Mayor James Dailey, most residents were unaware of the nature of the plant for three years. The *Sumter County Record*, a local newspaper, ran an article about the new industry entitled, "Unique New Industry Coming: New Use for Selma Chalk to Create Jobs." Many residents thought the new industry would be a brick factory (Bullard, 1990, p. 71).

Waste Management, Inc. claims that it is wrongfully seen as the villain. They believe that chemical waste produced by others is disposed of safely by them. But in the early 1980's, public attention was brought by the media to several violations of regulations by Waste Management. In 1983, the "Alabama Department of Environmental Management fined the company $150,000 for failing to complete a lining system. In 1984, they were fined for poor management procedures, PCB problems and inadequate ground water monitoring system at the site" (Bullard, 1990, p.72). In 1985, the EPA temporarily suspended shipments of superfund waste after test samples showed possible contamination of the aquifer.

Many people, including residents of the area around the Emelle site, worry that it is not possible to safely dispose of chemical waste, even if regulations are followed. More threat is posed if regulations are not followed. Since so much of the nation's chemical waste ends up in the Emelle facility, the local population feels, literally, "dumped on."

It is true that Waste Management benefits the community in some ways. Taxes paid by Waste Management provide a greater income to the county than residential property taxes. The tax benefits schools, libraries, ambulance service, and law enforcement. In addition, the company spends money purchasing goods and services in Sumter County. Four hundred new jobs were created, and these jobs pay more than the average local wage, although Wendall Paris, a Black civil rights activist in the area, claims only 50 residents from Emelle and Geiger (another primarily Black community in the area) work at the plant (Bullard, 1990, p.72). In addition, Waste Management tried in the early 1980's to improve its image with the community by sponsoring various community activities and offering tours of the site to explain its use and contributions to local charities.

One of Waste Management's spokespeople made clear its view that since toxic waste exists, something must be done with it. "If you wait until the best method for destroying hazardous wastes comes along, you will choke on your wastes," he said. "You have to go with the best technology you have now. You come up with the best standards you have and hope like heck they work."

But people in Sumter County are uneasy about how their lives will be affected by the toxins near their homes. They worry about the smells, the speed and number of trucks, occasional fires at the site, and about the discovery of traces of hazardous substances in test wells. When the county commissioners were asked if there was anything the company could do to reassure them about having the site in their community, they said, simply, "No."

CASE STUDY QUESTIONS

Names of people in group:

1. Look at the information about the make-up of the populations of the areas surrounding the two sites. The Commission on Racial Justice has already looked at this information. and has concluded that the sitings are based on "environmental racism."

a. What facts and statistics do you see that the Commission probably used to reach their conclusion?

b. Are there reasons that could be used to conclude that the sitings were not based on environmental racism.?

2. Imagine that you lived in one of these communities. Make a list of the reasons you would want the siting there. Then make a list of the reasons you wouldn't want it. Finally, if you could make the decision, what would you decide?

3. a. What did the protesters in Warren County lose in their attempt to block the siting?

b. What risks were they taking?

c. How do you think the protesters managed to get national attention? What would be worthwhile in doing such a protest campaign?

d. What did the protesters gain?

4. a. What contributions did Waste Management, Inc. make to the communities around the facility in Emelle, Alabama?

b. Why would these contributions be important to that community?

c. What problems are caused by the facility?

5. Imagine that a waste site is being considered for your community. What information would you need in order to form an opinion about the siting.

6. Some people believe that waste managers are not picking on the poor.
a. In what ways can you support this statement?

b. IHow can you show this statement to be is false? (Refer to the "Dumping on the Poor" article.)

7. Some people believe "environmental racism" influences what happens in our country, while others disagree. What would make one person believe one thing and another person believe something else? (Refer to "Dumping on the Poor.")

DUMPING ON THE POOR
America's dispossessed have lived for decades with toxic wastes and garbage. Now they are fighting back.*

The name Altgeld Gardens evokes images of brilliant flowers rampant in golden sunlight. But if you follow your nose into the black ghetto on the Far South Side of Chicago, it will lead you to a dilapidated housing project built atop a former landfill whose fetid odors still rise from the basements after more than 60 years. The plight of nearly 2,000 families is made worse by tons of pollutants from a nearby sludge plant, a steel mill, a paint company, a huge incinerator and an 80-ft.-high landfill. Only a few miles away is a lot that should be a playground. Instead it is a dump filled with 4-ft.-high mounds of trash, broken glass, rusty nails and construction debris.

In upstate New York, not far from the infamous Love Canal, you can follow your nose to Forest Glen, a trailer park settlement built on heaps of foul smelling hazardous waste that the Environmental Protection Agency says may contain as many as 150 toxic compounds. Under the streets of the densely populated semi-industrial section of Greenpoint, in Brooklyn, N.Y., the Mobil Corp. has begun recovering a sea of oil - 17 million gallons - that for decades has been leaking from underground storage tanks and pipelines.

Tens of thousands of impoverished people - mainly blacks and other minorities - living in the countless Altgeld Gardens and Forest Glens in the inner cities and rural pockets of the nation are the victims of what critics call environmental racism. The victimizers are mainly waste-management firms and local politicians hoping to attract revenues to their towns. They need cheap land where they can dispose of garbage and build air-contaminating incinerators. That all too often means land in poor areas with largely minority populations. And those people, burdened by drugs, poverty, crime, bad medical care and joblessness, have long been too powerless or apathetic to prevent their communities from becoming the repository of everybody else's detritus. The result, according to a landmark 1987 study by the United Church of Christ's Commission for Racial Justice, is that 3 of every 5 black and Hispanic Americans live in areas with uncontrolled toxic-waste sites. Many of the most notorious dumping grounds are located in the South. Among the worst is "cancer alley," a 75-mile stretch along the Mississippi River, from Baton Rouge to New Orleans, that is lined with oil refineries and

---

* Reprinted by permission of *Time Magazine*.

petrochemical plants. The alley's abnormally high cancer rate has prompted one health worker to call it a massive human experiment. A big mess in Chicago is the work of "fly dumpers," unlicensed truckers who collect filth from affluent neighborhoods and deposit it in vacant lots in stealthy forays at night.

Fearing that this appalling state of affairs can only get worse, the victims at last have begun to strike back. Often with the backing of ecological watchdog groups, grass-roots organizations are taking on the waste managers, using their publications and the law as their major weapons:

* In Altgeld Gardens, Hazel Johnson has organized a movement called People for Community Recovery, which has successfully crusaded against the neighborhood landfill.
* In Forest Glen, Terry and Kathy Freiermuth have shaken up the Federal Government, which has responded by promising buyout offers to residents as well as financial aid from the Federal Emergency Management Agency.
* In the blue-collar Pennsylvania village of Yukon (pop.1,100), Diana Steck is leading a protest organization of 600 members. Using roadblocks and other acts of civil disobedience as well as the legal system, the group is trying to force authorities to clean up six polluted lagoons that it suspects are killing livestock and causing cancer among the populace.

*In Passaic County, N.J., freeholders decided to build an incinerator in the city of Passaic on a site adjoining a hospital, whose occupants, according to a group led by Marge Gablehouse, would be the major beneficiaries of three tons of lead emissions a year. The protesters have succeeded in temporarily halting construction of the incinerator and hope to persuade Governor Jim Florio to cancel it.
* In California, Juana Gutierrez and her 400-member Mothers of East Los Angeles are fighting a proposed toxic waste incinerator slated for nearby Vernon, which every year would spew some threatening ash on their community. Why, she asks, should East Los Angeles, poverty-pocked and largely Hispanic, be subjected to this environmental atrocity? "Why not Beverly Hills or Bel Air?"

Waste managers deny they are picking on the poor. Some say it is simplistic to attribute the environmental problems of minority communities to racism, even though few challenge the evidence that the poor have more environmental dangers to cope with than do the wealthy. Blacks, Hispanics, and Asians have often inherited hazards by moving into older sectors of cities, where decrepit factories and other facilities were built long before anyone worried about pollution.

Government officials have often spurned complaints from low-income residents about the hazards posed by landfills and incinerators. To bolster their credibility, waste management firms have hired dozens of former officials of the federal Environmental Protection Agency and private conservation groups. Former EPA chief William Ruckelshaus, for example, is now chairman of Browning-Ferris industries, one of the largest waste-treatment outfits in the country. In October, after the mostly Hispanic residents of Azusa, Calif., complained that expansion of a BFI-operated landfill would poison the groundwater, BFI offered to invest $20 million to clean up the contamination. The expansion was approved.

Sometimes dumping on the poor is a consequence of self-exploitation. In their search for new disposal sites, waste managers have discovered fertile ground on American Indian reservations, which are considered soverign entities not subject to local or state environmental restrictions and whose residents are perhaps the poorest of the nation's poor.

Typical is the dilemma of California's Campo tribe of Mission Indians whose land lies 68 miles east of San Diego. Leaders of the community of 250 people are negotiating with a waste-disposal company to build a landfill and recycling plant that would be fed up to 3,000 tons of garbage a day from San Diego county. This would be a boon for the county, which is running out of landfill. For the Indians, the project would bring jobs and "millions" in income, says tribal EPA chairman Michael Connelly-Misquish.

Connelly-Misquish claims the tribe would write the health and safety codes at least equal to those required by California's environmental agencies. But ranchers and farmers near the reservation are not so sure. Concerned that the dump and incinerator would contaminate the region, they have asked the state legislature to make all California reservations subject to state environmental regulations. If they succeed, the waste companies can expect to come up against tougher environmental rules at all other reservations.

The Campos opponents are practicing a strategy that wealthier neighbourhoods use regularly with considerable success. That strategy is encompased in a now familiar slogan: NIMBY (not in my backyard). With the help of money and political power, such groups fight off unwanted facilities like half-way houses for recovering addicts, prisons and incinerators, which then almost invariably find homes that can least afford to resist them. In well-to-do Greenwich, Conn., for example, singer Diana Ross and 12 fellow residents launched a NIMBY campaign to prevent the state from building a $6 million-plus truck-weighing station on an interstate highway that runs near their homes.

For the nation's newly energized grass-roots activists however, NIMBY is a fruitless answer. Their's is a challenge that the entire US has yet to confront: it's spelled NIABY --

not in anyone's backyard.

Where, then? The activists aren't terribly helpful in answering that question. They are, however, eloquent in arguing that the nation's monumental waste-disposal problem can no longer be solved by transferring toxic trash from a privileged neighborhood to a less fortunate one; that even state-of-the-art incinerators are costly and inefficient; that a nation that produces 160 million tons of residential and commercial solid wastes every year must find a way to deal with them without destroying its communities; and that, finally, pollution kills people. The rate of lung cancer among young urban blacks is significantly higher than the national average. Some experts persuasively suggest that this dismaying statistic may be attributable not so much to rotten ghetto air as to the fact that young blacks smoke too much.

Regardless of the merits of that debate, one thing is clear: the war to protect the environment has opened a new front in gritty ghetto streets and downtrodden rural backwaters where the endangered species is the poor.
--Reported by Jerome McDowell/Los Angeles, Elizabeth Taylor/Chicago, and Liza H. Towda/New York

CHAPTER FOUR

MAKING A DIFFERENCE

OVERVIEW FOR TEACHERS

Chapter Four reflects three underlying beliefs. First, taking action is an important aspect of learning about complex contemporary issues. Second, individuals and small groups of citizens working together can make a difference. And third, taking effective action offers opportunities to learn and develop a wide variety of skills.

This chapter includes articles about individuals taking action in Activity 26 and a plan for students' action projects in Activity 27 Through the readings and projects, students can:

- learn about change and participation in society
- learn about different roles in our communities and how people and institutions operate together
- develop a plan, delegate tasks, and evaluate and reformulate the plan when necessary
- develop research, communication, leadership, and group participation skills
- create and present reports.

As mentioned in the general introduction, taking action can make people feel empowered as they learn about important issues in society. Such issues are often quite complex and overwhelming, and students can have difficulty seeing a role for themselves. For this reason, we included several activities earlier in the curriculum through which students could participate in some aspect of the larger trash disposal issue, such as designing less wasteful packaging and reducing their own lunch waste. In this section, students have the opportunity to think about trash disposal comprehensively, select an aspect that is important to them, and try to make a difference.

Encouraging students to be active in their communities by guiding them through the process of researching issues and selecting actions helps prepare them to be contributing, active adults in the future. But taking action can be important in the present as well, for young people gain a sense of their own worth and ability to control their world when they join others who are trying to find solutions to problems.

Through the first three chapters of this curriculum, students have learned that the volume of trash in the United States is a concern, and that each existing disposal method has advantages and disadvantages, that there are ways to reduce the amount of waste (and hazardous waste) produced. They have also learned that social conditions and political processes influence what is done with trash. The next activities allow students the opportunity to develop their own influence.

Activity 26

INDIVIDUALS MAKING A DIFFERENCE

Objective

Before students take action themselves, it is helpful for them to think about how and why others became active. In this activity, students will learn about people who worked to clean up toxins or make other environmental changes in their communities or through their businesses. They will also determine what factors contribute to the transformation of everyday citizens into activists.

Suggested Previous Activities

15. Packaging
18. An Example from Industry: 3M's 3P's
19. Hazardous Waste In Schools
24. Reservations About Waste
25. Uncovering Environmental Injustice

Time

homework assignment prior to class (reading 2 articles)
45 minutes for discussion and video

Vocabulary

activist   hazardous
boycott   toxin
emissions

Materials

1 copy of Handout 51, "Saying 'No': Boycotts & Consumer Action," for each student
1 copy of Handout 52, "Earthen Renewal Begins at Home," for each student
1 copy of Handout 53, "Toxic Avengers: Teenage Mutant Crusaders Against Pollution," for each student
1 copy of Handout 54, "The Environmental Scoop on Ben & Jerry's," for each student
1 copy of Handout 55 "Statements from Patagonia" for each student
videotape, *Fire in Their Hearts*, available from Boston Area Educators for Social Responsibility (617) 492-8820
pencils and paper

Activity Procedure

1. Have the students read the articles for homework. They can each read all of the articles, or each group of students could receive a set of the articles. In the latter arrangement, each student could take responsibility for one or two articles, then share what they learned with their group in jigsaw fashion. Review with students what they learned through any of the previous suggested activities listed above.

2. In class, have students work in small groups to answer the following questions and any other questions that may arise.. Answers from small groups can be written and handed in or used as a basis for discussion with the whole class.

> * What boycotts have you heard of, and why are the products being boycotted?
> * In what ways did people in the article "Earthen Renewal..." become active at home"?
> * Are people more likely to become active about issues in their own communities? Why or why not?
> * In what ways do you think Ben & Jerry's might be different from other ice cream companies?
> * What do people or businesses gain and lose in making change?
> * How similar is your shopping philosophy to Patagonia's production philosphy?
> * Which story was most interesting to you, and why?
> * If you could, which activist from the articles would you like to interview and what questions would you ask?
> * Do you know of groups in your own communities that are trying to solve environmental problems?

3. Show the section of the videotape *Fire in Their Hearts* that tells the story of Lois Gibb, the leader of the Love Canal protests in Love Canal, NY.

4. Encourage a class discussion about the video.

> * What kind of person was Lois Gibb before she took a leadership role in her community?
> * How was she different after she took on that role? What qualities do you see in her as a leader?
> * What do you think made her take responsibility for leading her neighbors in getting retribution for the destruction of their community?
> * What change took place in Lois Gibb's attitude toward the government and why?
> * Why did the community feel it had to resort to taking hostages? Do you think that such extreme measures are sometimes necessary? Why or why not?
> * Why do you think Lois Gibb didn't stop being an activist once she had left Love Canal?
> * What conclusions can you draw about
>   - what makes a person become an activist.
>   - what kind of person might become an activist.
>   - what qualities are necessary to be an activist.
> (In answering these questions, students might find it helpful to generate a list of words that describe Lois Gibb.)

Practical Actions for Conscious Living

🚫 Saying "No": Boycotts & Consumer Action

THE HARVEST WAS BAD AND famine was predicted in Ireland during the summer of 1880. Charles Stewart Parnell, president of the tenants' rights group, the Irish Land League, told landlord Captain Charles Boycott that he must reduce rents by 25%. The tenants, Parnell told the captain, simply could not pay. Boycott refused the request.

The enraged tenants followed Parnell's advice, denying their labor to Captain Boycott. By autumn, nearly every European newspaper was using the word "Boycott" to describe the tenant action.

Tea Parties and Other Revolts

Similar actions had occurred before this, under other names: the Boston Tea Party of 1773; the frequent refusal of 19th century American farmers to patronize certain railroads until they lowered their rates. In 1885 alone there were 196 recognized boycotts by American labor groups, according to the Encyclopedia Americana.

In the 1950s and '60s, the civil rights movement used boycotts very effectively against businesses that practiced racial discrimination. In 1980 the United States called for a boycott of the Moscow summer Olympics to protest the Soviet invasion of Afghanistan.

All had one thing in common—individuals uniting to create a group with enough power to pressure for economic, social or political change. Every one of these groups, however, began with one individual's actions, example or leadership.

Reprinted by permission of *One Person's Impact*

Boycott
Continued from first page

How An Individual Consumer Boycott Works:

1. As a consumer you wield considerable economic clout. The decisions you make almost every day concerning how to spend your money either support or hurt companies competing for your dollars. If you choose not to purchase products because of a company's political affiliations, or environmental record, you are boycotting that company.

2. Even if there is not, to your knowledge, an organized boycott of a product or practice that offends you, don't assume you're alone. Put your money where your ideals are (even if that brand is on sale this week!) and don't buy it.

Examples of Effective Boycotts

Tuna—For some time now there has been concern over the number of dolphins and other sea animals killed during fishing for tuna with miles-long "drift nets." Due in part to consumer boycotts of tuna, the Dolphin Protection Consumer Information Act of 1989 was introduced, which would require that tuna labels state the fishing method and whether that method is harmful to dolphins.
[NOTE: As this article was going to press, the makers of Starkist Tuna announced that they would no longer buy or sell tuna caught in nets with dolphins. Bumble Bee and Chicken of the Sea instantly followed suit. Cans with "dolphin-safe" labels should be appearing on grocery shelves in the next few months.]
Icelandic Fish—Greenpeace recently ended a two year boycott of Icelandic fish, which, according to Greenpeace, cost the country $50 million and was "waged as an economic lever to halt the country's whaling practices." Iceland's Minister of Fisheries has promised that "research" whaling ships will not sail this year. (From *Boycott Action News*)
Plastic Egg Cartons—According to *Resource Recycling* magazine, consumer demand reversed a trend toward plastic egg cartons. Led by retailers unwilling to incur "ill will" from consumers, the Eastern Canadian market turned to pulp (paper) egg cartons in one year.

3. Then go one step further. Write a letter. Yours could be the first, or the thousand and first message the company president needs to convince him or her that change is necessary.
4. Don't neglect those who make changes or who do the right thing in the first place. Buy their products. Write letters of thanks. We all need encouragement.

How to Write an Effective Letter

- Start with the customer relations department and obtain the correct name and address. Usually this will be printed on the product's packaging. If not, check a reference book (see Additional Resources).
- Use the business letter format.
- Be brief and to the point. State your

> Dear Sir or Madam:
> It's an extremely busy time of year, but I am so offended by one of your products I'm making the time to write you. Last night I noticed a "new" product of yours—[Product Name]—a plastic container approximately one-third filled with dry pancake mix, to which one "simply adds water" and shakes. Then the entire plastic container can be tossed. How simple, how convenient, how wasteful!
> Given the growing awareness of the solid waste problem in this country, the commendable move towards recycling, reuse and source reduction, I can't imagine why you've chosen to introduce this overpackaged product that simply caters to the laziest among the population.
> Frankly, I'm outraged that you've done this. I will no longer buy your products, will urge others to do the same, and will do whatever I can to publicize this grossly irresponsible behavior on your part. Shame on you.
>
> Sincerely,
> A concerned citizen

objections, your course of action and what you believe should be done. Typewritten letters are easier to read than handwritten ones.
- You should receive a reply. If you do not get an answer, or you are not satisfied with the one you get, you may want to write directly to the company president or another executive in charge of public relations, marketing, etc. Or, write to them all!
Sample letters are reproduced on this page.

In Conclusion
In September 1880, Captain Boycott was forced to hire outside workers to har-

vest his crops and soldiers to guard them. He left Ireland the same year. The next year, under William Ewart Gladstone's Land Act of 1881, conditions eased and fair rent tribunals were instituted.
Then, as now, the boycott was the first step in bringing an intolerable situation to the attention of those responsible, and informing them that changes must be made.

—Paula Haney

> Dear concerned citizen:
> Thank you for contacting us.
> [Our Company] understands and commends the concern that consumers, such as yourself, have for the environment. The majority of products that we produce are packaged in biodegradable or recyclable containers. [Product Name] is packaged in a recyclable plastic bottle.
> Choosing a packaging material is a complex issue. Many factors such as safety, shelf life, the ability to prevent moisture loss or absorption, package strength, and consumer preference determine which packaging is most suitable.
> Your concerns are being shared with the appropriate [Company] personnel. We appreciate that you have taken the time to express your point of view.
>
> Sincerely,
> Consumer Affairs

►Additional Resources:

- Several public library reference books list products and companies that make them. Among them are the *World Almanac* under "Consumer Information," and *Brandnames:Who Owns What*, by Facts on File.
- *Standard and Poor's Register of Corporations, Directors and Executives*, reference section, public library. Provides more detailed information such as officers of the company, background information, subsidiaries, and much more.
- *Building Economic Alternatives/Boycott Action News*, Coop America, 2100 M Street,Suite 403, Washington, DC 20063, 202-223-1881 Features latest boycotts, who they were called by and why, addresses, letter reprints, etc.
- *National Boycott Newsletter*, 6506 28th Ave. NE, Seattle, WA 98115. In-depth reports on boycotts, with references and addresses. ◊

impact fact ► For faster composting of kitchen waste, chop potato peels, broccoli stems, egg shells, and so on, in your blender or food processor. In the yard, leaves and light brush can be shredded by running the lawnmower over them a couple of times.

EARTHEN RENEWAL BEGINS AT HOME*

Fearing pollution, local activists are putting their feet down and fighting the power

While celebrities and scientists worry about the ozone, Jose´ Morales of New York city has a more local concern: residential pollution. "Who's going to live closer to the factory," he asks rhetorically, "the factory owner or the worker?" With toxins lurking right next door these days, Americans have begun a movement to re-green their own back yards.

For years street gutters in the Williamsburg section of Brooklyn had carried caustic-smelling effluents from an adhesives factory. Under the tutelage of high school science teacher Morales, 28, students took samples of the runoff back to their lab and found them flammable and poisonous. When they also discovered a radioactive and hazardous wastes store-house in their neighborhood, they banded together as the **TOXIC AVENGERS** and organized 100 neighbors for a protest march. Coming next: a town meeting to spur further community action.

In California, the **MOTHERS OF EAST LOS ANGELES** sued the state to halt construction of an incinerator in their Latino neighborhood. "It would have emitted chemicals bad for kids and pregnant ladies," says Juana Gutierrez, 57, the mother of nine, who co-founded the group. Adds Aurora Castillo, a sexagenarian campaigner, "We used to say we can't fight the government. But we *are* the government."

In a Chicano section of Albuquerque, New Mexico, **JEANNE GUANA**, 43, analyzed a 1985 health survey and found that nearly all of her neighbors suffered rashes, respiratory problems or both. Discovering that sawdust from a nearby particle-board factory was tainted with formaldehyde, the community roused regulatory agencies to force the factory to limit its emmisions.

Rural communities, too, are digging in. A fourth-grade class in Savannah, MO, raised newts, turtles and catfish. But at the Missouri River tributary where they planned to release the 12 indigenous species, the children found the shore littered and the water polluted with soil runoff from an adjoining farm. They picked up the trash, planted hundreds of willows to hold the river banks and persuaded the farmer to stop plowing nearby. Teacher **SHERRI STRATING**, 41, applauds her students. "They really put some challenges to me. One of my little girls says, 'We're the adults of the

---

* Reprinted by permission of *People Magazine*

future. If nothing is done about this now, we're not going to have anything later.' "

Pat Bryant, 43, a Louisianian who heads the **GULF COAST TENANTS ORGANIZATION**, estimates that a half billion pounds of toxic chemicals a year are spewed over a 75 mile strip between Baton Rouge and New Orleans. An 11-day march down "Cancer Alley" prodded 1989 legislation to control emissions. Says Bryant: "The poisoned are the only ones with the power to stop the poisoning."

When the L.A. Board of Ed. rejected a recycling plan from North Hollywood High senior **ALLEN GRAVES II**, 18, he was incensed: "A planet is being stolen from 10 million teenagers." After calls, letters and an impassioned plea, he got the board to reverse itself and for his efforts won the President's Environmental Youth Award. "Kids are always told what not to do," says Graves. "But when you tell them they can make a difference, you've created an incredible amount of energy."

Toxic Avengers
Teenage Mutant Crusaders Against Pollution

Linda R. Prout

At a community center in the Williamsburg section of Brooklyn, New York, members of a panel are talking about sitting-in, lying down, linking arms, and chaining themselves to fences. The enraptured listeners are getting lessons in civil disobedience, tactics practiced by civil rights and anti-war demonstrators years before most members of this young audience were born.

But this is no group of starry-eyed youngsters, marveling at the old war tales of elder activists. These are the Toxic Avengers, Latino and African American high school and college students who have rallied and marched to rid their neighborhood of hazardous materials. Now they are contemplating their next move.

One of the panelists, Gilbert Martinez of the Latin Caucus of ACT-UP, the AIDS Coalition to Unleash Power, recounts how police roughed up and arrested scores of people who sat-in at a Chicago hospital earlier this year. Benny Vasquez, one of the co-leaders of the TAs, as the Toxic Avengers are called for short, worries that an arrest might remain on his record. Another TA is afraid of how her parents would react.

"I think civil disobedience is for the future," says Rosa Rivera, 20, a student at Kingsboro Community College in Brooklyn and a former leader of the Avengers. "We're not ready yet. But we need the information for when we are ready."

The Toxic Avengers might not be ready for civil disobedience now, but when they are, they will know their target. For two years now, these young environmental activists have waged a campaign against the Radiac Research Corporation, a company which stores and transports toxic waste, including low-level radioactive and flammable materials. They have plastered the neighborhood with posters and handed out leaflets detailing

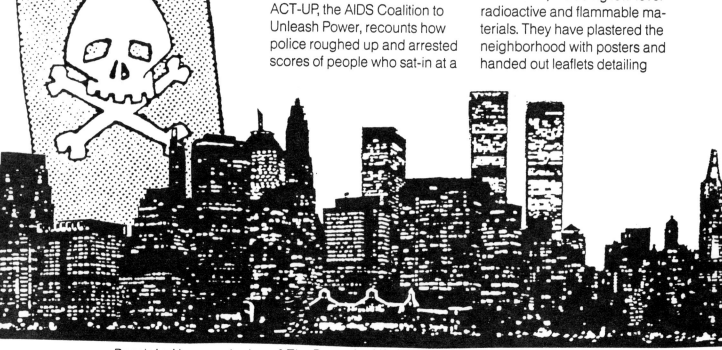

Reprinted by permission of The Panos Institute

what they see as the risks of having Radiac in their community. And probably most impressive, the TAs have organized three demonstrations against the company, including one in March of 1990, co-sponsored by the New York Public Interest Research Group, that attracted 200 marchers.

"This company is a hazard to the community and we want the community to be aware that it can do something about it," explained Rivera.

The Toxic Avengers' community is Williamsburg, a working-class neighborhood that shares a peninsula in northeast Brooklyn with the community of Greenpoint. The neighborhood has attracted immigrants since 1903, when the Williamsburg Bridge opened and Jews from New York's overcrowded Lower East Side flowed into the area. Today, Williamsburg is home to about 35,000 people, mostly Latinos from the Caribbean and Central America. It is also the home of some of New York's most serious polluters. Last year, the Community Environmental Health Center at Hunter College in Manhattan profiled 28 facilities in the Williamsburg-Greenpoint area that stored large quantities of toxic waste. Eleven facilities stored substances that the Environmental Protection Agency (EPA) termed "extremely dangerous" in quantities ex-

ceeding EPA recommendations, the report said.

Add to these worries the pollutants breathed in daily by simply living in a neighborhood crisscrossed by the Williamsburg Bridge and the elevated Brooklyn-Queens Expressway, two roadways almost constantly jammed with cars spewing toxic fumes.

Radiac, which has been doing business for 21 years next to a public school playground, was an obvious target. Although the state's Department of Environmental Conservation maintains that Radiac has a good safety record, residents have for years criticized the location of the facility. But it took a dozen teenagers, young people of color, to actively challenge the company. The Toxic Avengers were well aware that as minorities, they were likely to be the victims of environmental racism.

"Most toxic sites are in poorer, mostly minority communities," offered Vasquez, 15.

Adds Rivera: "The government could do something about Radiac. But they figure that just minorities live around here and we're not going to do anything about it."

The Toxic Avengers work out of a community center called El Puente, which means "the bridge" in Spanish. The name stands for more than its location in the shadow of the Williamsburg. It symbolizes the center's mission, bringing this community together through positive, productive endeavors. The Toxic Avengers began here in 1988, evolving out of a project for a class in environmental sciences for high school dropouts study-

ing for their Graduate Education Diplomas (GED). Jose Morales, who taught the course, sent his students out into the streets of their neighborhood. They found an empty lot filled with metal drums from which toxic liquids were oozing. Environmental agencies were called in and the lot was cleaned up.

But Morales and two of his students decided there was more work to be done. They invited Rosa Rivera, an active El Puente member, to join them and the Toxic Avengers were born. Morales named the group after a movie character, a teenager who fell into a tub of radioactive gook and emerged a mutant crusader against polluters.

Eventually, other teenagers became interested, too. Over the last two years, the group has grown to twelve. The nine high school and three college students, ranging in age from 15 to 20, meet once a week. The group has eschewed traditional hierarchy; instead of presidents, there are two "co-leaders" and the positions are rotated every six months. Actions are decided by majority rule. Morales, the group's advisor, attends meetings but has no vote.

Belonging to the organization has given many of them their first taste of empowerment.

"I was very timid, very shy," admitted Rivera. "But since I've been a Toxic Avenger, I've sort of bloomed."

In addition to battling Radiac, the Toxic Avengers' current agenda includes starting a recycling program at El Puente. If that effort is successful, the TAs will go to a different building each week and teach residents about recycling. They also will spend a good deal of time holding workshops to convince community members, especially other young people, that combating pollution should be a priority just like fighting drugs and crime.

And the Toxic Avengers have not confined their message to their community. On the contrary, they have become "environmental ambassadors," traveling to conferences and youth meetings, and conducting workshops around New York and in other parts of the country. In the process they have become role models, especially for inner-city youth.

"I think when young people see us, they feel empowered because we feel empowered," said Vasquez. "I tell them that every youth has the right to think for himself and to say and do what he feels is right."

What has struck the TAs most during their travels is the absence of people of color from the environmental movement. At a conference on the environment in Illinois, only five percent of the participants were Latino or African American, Rosa Rivera recalled. Racial and class distinctions were even more obvious at an Earth Day conference in New York last April. Rivera said the Toxic Avengers were the only Latinos in attendance. She described most of the other participants as "older, more powerful Caucasians with business cards."

To fight this kind of exclusion, the Toxic Avengers now print all their literature in English and Spanish. Ironically, they are attracting just as much attention from mainstream groups.

"I think they're great," said Steve Kretzmann of Greenpeace USA, an environmental group that has been criticized for its "Whiteness." "By empowering themselves, they can empower their community."

And recognition has come from other quarters. The Toxic Avengers have received public service awards from the Citizens' Committee for the City of New York and The New York Daily News. They have been invited to address the New York State Legislature in Albany early next year.

"Groups like the Toxic Avengers can make a significant difference," said Assemblyman Joseph Lentol (Democrat, NY), who represents Williamsburg and Greenpoint.

The Toxic Avengers already know this. "We're the present and the future," said Benny Vasquez. "If we don't fight for the environment, the next generation will suffer." ∎

LINDA R. PROUT is a journalist specializing in the Caribbean and the Caribbean diaspora.

THE ENVIRONMENTAL SCOOP ON BEN & JERRY'S*

How Could Ice Cream Harm the Environment?

At Ben and Jerry's Ice Cream company they realize that all businesses, even an ice cream business, have a major impact on the environment. Some of the environmental impacts include the following:

• the office workers generate a lot of paper
• cardboard is used in the packaging
• eggs which are used in the making ice cream come in plastic containers
• there is always ice cream spillage
• and making ice cream creates waste water.

Ben and Jerry's has implemented a number of programs to lessen its environmental impact. The overall goals of Ben and Jerry's are to responsibly manage their solid waste, conserve energy and resources, and develop environmentally beneficial community programs.

How Do They Reach Their Goals?

Ben and Jerry's created the position of Manger of Natural Resources Use to work toward their goals. In addition, they have a volunteer "Green Team" that helps improve the company's environmental impact in many areas of their operation.

Everyone recycles.

The volunteer Green Team has made sure that all office employees have recycling containers for all kinds of paper (white, mixed, glossy, and newspaper) at their desks.

Cardboard recycling makes money.

Ben and Jerry's generates about 18,000 pounds of cardboard per **week!** The factory has its own cardboard balers. The bales are sent to a paper broker every week. Approximately every ten days 5,000 new boxes are made by a company in Canada from the recycled cardboard. Ben and Jerry's saves $17,400 in annual hauling costs, since less trash is thrown out which they would have had to pay to have hauled away.

Plastics can also be recycled.

The eggs used in the ice cream are packed in plastic pails. Ben and Jerry' s uses about 1,000 plastic pails per week. Ben and Jerry's works with the Vermont Republic Industries to have the plastic pails cleaned, shredded and sent to a plastic manufacturer for recycling. Some buckets are also washed and reused. Ben and Jerry's saves $235,000 per year by not sending these plastic pails to the landfill. Other kinds of plastic are recycled as well. Production workers recycle plastic bags and stretch wrap that come in from suppliers' packaging.

They have reduced packaging.

Ben and Jerry's switched to 100% recycled paperboard for their Brownie Bar and Peace Pop boxes. They no longer use a tray in the Brownie Bar packaging. What about the pint containers that the ice cream is packaged in? Ben and Jerry have researched this problem and found that there are five recycling plants in the United States that specialize in poly-coated paperboard but they have not figured out how to get the empty cartons to those recycling plants. If you have any idea write to Ben and Jerry and let them know. Write to: Green Team Resources, Ben and Jerry's, P.O. Box 240, Waterbury, VT 05676.

Ben and Jerry's sends ice cream to pigs?

What happens to all the ice cream that spills and can not be sold? About 500 to 700 gallons of ice cream spillage each day are sent to pig farms. Leftovers, such as the cherry juice from Cherry Garcia ice cream, are also sent to the pigs.

Waste water is treated the "greenhouse way."

Making their ice cream creates a lot of natural waste. The natural waste leaves the factory and goes to the company's waste water pre-treatment lagoon and then to the greenhouse, which is powered by solar energy. The greenhouse creates an environment where plants, algae and microbes feed on the waste water, thereby breaking the waste water down to pure water and a non-hazardous byproduct. The plants, algae, and microbes grow while the waste water is treated. Ben and Jerry's greenhouse system is the first of its kind ever used to process dairy waste.

Merry Mulching

Vermont's Merry Mulch program sets up centers for chipping Christmas trees after the holiday is over. Any Vermont resident who brings a tree to Merry Mulch center gets a free Ben and Jerry's cone coupon along with as much mulch as they want.

The Green Flea

Employees of Ben and Jerry's donate truckloads of items to sell at a day-long flea market. All proceeds from the flea market go to the "Environmental Federation of America/Earth Share." This is an umbrella group for several environmental organizations.

*Material adapted from "Ben and Jerry's 1991 Annual Report," and "Ben & Jerry's Environmental Care" brochure, Ben and Jerry's Homemade, Inc., Waterbury, VT 05676.

STATEMENTS FROM PATAGONIA

The outdoor clothing company, Patagonia, recently did an assessment of their company's environmental impact. They looked at waste from the manufacturing processes, the impact of growing cotton, the effects of the ways in which sheep are raised for wool, and much more. They decided to change their company's philosophy. Below are a few excerpts from their "Simple Gifts" statement of philosophy.

...Everything we make pollutes. Polyester, because it's made from petroleum, is an obvious villain, but cotton and wool are not any better. To kill the boll weevil, cotton is sprayed with pesticides so poisonous they gradually render cotton fields barren; cotton fabric is often treated with formaldehyde. Wool relies on large flocks of sheep that denude fragile, arid areas of the earth. We are pursuing other alternatives, like...using "organic" cotton....Another conclusion we have reached is simple: we need to use fewer materials.

As a society, we've always assumed that growth is both positive and inevitable. When our economies sour, as they always do, we simply look for new consumers and new resources. We hunt new export markets and new Third World sources for raw materials. But Third World resources are close to exhaustion and many world economies,...are no longer viable dumping grounds for our manufactured goods....This is a vital concept to grasp: growth has always been assumed to be good in American culture; bigger is better. But those days are over.

We don't see [the nineties' recession] as a temporary recession. Instead, [it is] possibly the end of economies based on consumerism....In the United States alone, it is clear that people are no longer interested in shopping as entertainment....

...we are limiting Patagonia's growth in the United States with the eventual goal of halting growth altogether. We dropped 30% of our clothing line....Last fall you had a choice of five ski pants, now you may choose between two....two styles of ski pants are all that anyone needs....You will see fewer color and style choices throughout the catalog.

We think that the future of clothing will be less is more, a few good clothes that will last a long time.

Activity 27

TAKING ACTION

Objective

Through this activity students will learn about principles of democratic participation and ways to take action on issues that are important to them. Students will research a problem related to trash and its potential solutions; they will then plan and carry out an action project related to that issue.

Note

These suggestions will allow you to set up projects for independent work for small groups of students or to create class projects. These steps are not set in concrete. Depending on the topic and your students, you may want to change the order, skip steps, or insert others. We have organized this activity into five parts: Focusing the Issue, Deciding on an Action Project, Developing a Strategy, Working on the Issue/Conducting the Project, and Final Presentations. Making a difference in one's community is rarely simple or quick. It will be important for students to have some perspective and do a lot of preparation so they will experience success.

(Much of this section was adapted from *Making History: A Social Studies Curriculum in the Participation Series,* Boston Area Educators for Social Responsibility, Cambridge, MA, 1984.)

Time

1-3 class sessions for preparation
time for doing projects will vary

Materials

1 copy of Handout 56, "A Range of Actions and Projects" for each group of students
1 copy of Handout 57, "How to Get Things Done" for each student

Activity Procedures - Part A: Focusing the Issue

The general topic of trash includes an enormous array of subtopics. Have students brainstorm a list of particular issues. They might name, for example, waste at school or waste in a particular part of town. They could focus on a particular kind of business, such as restaurants or gas stations. They could focus on a specific kind of waste, perhaps diapers, paper, batteries, or glass. Students might make lists of local, regional, national, and global issues. (It's usually much easier to work on a project that has at least some local component.) Have students select something they want to work on and, if they are working in groups, the one or two people they will work with.

Activity Procedures - Part B: Deciding on an Action Project

Students may have focused on an aspect of the trash issue that leads easily to a choice of one or another kind of project. For example, they may have named as their concern that their school buys too much newly made paper. To deal with that, they might find out who makes paper purchasing decisions, invite them to class, and ask them to change the school's brand of paper.

On the other hand, a particular trash issue could be acted on in various ways. Decreasing paper waste in school might lead to a recycling project, changes in how school lunches are provided, doing away with paper passes for the hallway, or even changing some of the practices in the art room.

Have students brainstorm different kinds of action projects that might be appropriate for their topic. Handout 56 offers a wide variety of options. If students have chosen a topic but are not sure what action project is appropriate, they may need to read some articles or interview one or more people to get ideas.

Activity Procedures - Part C: Developing a Strategy

Now that students have chosen an issue to focus on and have thought of various ways to approach that issue, they will be ready to understand the context in which they mean to make a change and develop an action plan. They will be able to understand more clearly the complex interrelationships among individuals and groups who influence decisions and the best ways to approach others to create change. To do this effectively they will go through the following exercises:

Exercise 1. Identify key people -- who has influence, who has responsibility, and who is affected -- so that they know who they will need to work with on the issue.

Exercise 2. Determine the obstacles to accomplishing their goal, as well as the resources available to overcome these obstacles.

Exercise 3. Consider ways they might build support.

Exercise 4. Outline a plan and a time line, and delegate tasks and roles.

Exercise 5. Practice, through role-playing, the communication skills they will need to use.

Organizing to effect change may be new for students. We have included Handout 57 which outlines some basic organizing principles. The suggestions we have made are different in many respects from the traditional organizing strategies of political campaigns or political action groups. We have articulated a strategy based on understanding other perspectives and points of view, finding common ground, and acting in ways that demonstrate respect for all those involved. We have chosen this strategy because it continually seeks ways to work cooperatively and because it minimizes antagonism and polarized positions -- two obstacles to meaningful change.

As students have learned about the trash issue, they have seen that the topic is complex. There are many different views as to what is most important, what needs to happen, whose responsibility it is, etc. They have also seen that changing practices, whether personal habits or community, national, or corporate policies, is neither simple nor fast. Planning ahead, assessing resources, and looking for allies, are all part of effective organizing.

Exercise 1: Who Has Power? Who Has Responsibility? Who Is Affected?

This exercise will allow students to map out the relationships of those involved in the issue so they may see effective entry points for their efforts as well as the people they might want to approach for support. It will also help them realize that there are many people involved in any issue, and that each has his or her own interests and positions.

Begin by brainstorming a list of all the people who are either affected by the issue or who have decision-making responsibility. Have the students divide this list into those who have direct responsibility for making the decision, those who have influence on the decision-makers, and those who are affected by the decisions. The last two categories may overlap.

Next, have students diagram the relationships among these people and groups, using arrows to show the direction of influence. For example, if students were attempting to promote legislation for a bottle and can bill in their state, their diagram might look like the following:

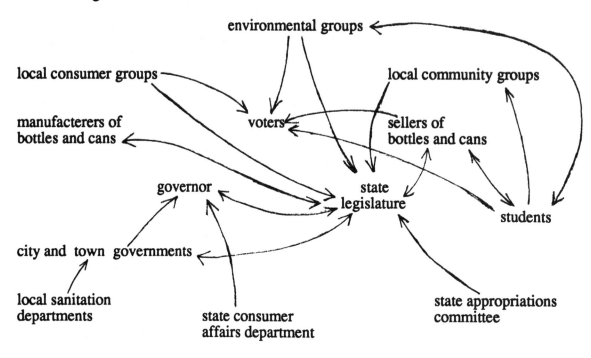

Put the diagram your students develop onto a piece of poster board or newsprint and leave it up for students to refer to throughout the planning process. Awareness of the lines of power and responsibility will make possible a more effective plan of action.

Exercise 2: What Are the Obstacles? What Are Our Resources?

The purpose of this activity is to help the class anticipate the obstacles they will confront and identify the available resources for overcoming them. You may want to begin by exploring the concept of obstacles in general and then brainstorm with your students the specific obstacles -- rules, attitudes, special interests, or other difficulties -- they might encounter in their efforts to create change. In the case of a bottle bill, for example, the obstacles might include the higher costs to local businesses, increased taxes to citizens and apathy among voters. Write this list on the side of the chalkboard.

Next, describe the concept of resources -- supportive individuals or groups, talents of the students, services of the school, such as a print shop or Xerox machine, etc.-- and on the other side of the chalkboard, brainstorm a list of resources available to the class. Have your students look at both lists and identify those resources that seem useful in overcoming specific obstacles. Draw arrows from the resources to those obstacles. Have the students look again at the obstacle list and, anticipating the future, ask them to consider additional resources to overcome them. In this way, students can begin to understand what strengths they have, at what points they might have easier access to decision-makers, and what actions might be most important to achieving their goals.

Exercise 3: How Do We Build Support for Change ?

Before you actually begin to brainstorm and plan action projects, we would suggest that you spend some time discussing how people feel when they are asked to change something and how people can be approached so that they are the most supportive and least defensive. This discussion may influence the students' choice of action by pointing them in the direction of those actions where less opposition is anticipated.

You may want to begin by sharing with students that because change can be risky and some people are satisfied with leaving things the way they are. If we, as a class, are asking people to change, we should have some understanding of what that might mean to them and how we might best approach them.

Ask students to think of a time when someone either asked or insisted that they change something. This could be an experience with a friend, their family, or in school. Discuss with them whether it was difficult to change, and what made it difficult.if, they resented it. You may want to make a list of those things that made it easier to change and those that made it harder. The following questions may stimulate ideas:

* Did you see the benefits of the change quickly? slowly? ever?
* What was it in the way that you were approached that made it either easier or harder for you to change?
* What made you want to dig in your heels?
* What made you want to change willingly?

Out of this list will come a sense that when people are treated with respect, listened to, and understood, it is easier for them to change. When they are threatened, people often resist and resent the change, even when they recognize its benefits. You may want to have students record some of their thoughts about this discussion in their journals.

Next, distribute Handout 57 "How To Get Things Done," and hold a class discussion using the following questions:

*How consistent are these principals with what you experienced during your own changes?
*How consistent are these principles with what often takes place when people try to create change?
*In what ways do you agree with these principles?
*In what ways do you disagree?

Exercise 4: Who Will Do What, When?

Give students paper and ask them to list all the tasks they will have to do on their projects. It will probably be helpful to give them scrap paper first and then large sheets of newsprint or computer paper or they could create a table in a computer if one is accessible. It is difficult to rememeber or imagine all of the tasks in order on the first try.

When they have written all the tasks they can think of in the order they will be done, have students write their name or initials next to the tasks they will be responsible for and an estimated date when they will do the task. If students create this list in columns a fourth column might be useful for a date or check mark when the task is accomplished. The following chart is an example of what could be done.

Tasks	Who?	By when?	Done

Encourage students to view their charts as guides and reminders. Some tasks will take more time than estimated, others will take less. They should be used to help students stay on course and watch their own progress, not to make them feel frustrated or behind schedule.

Exercise 5:

Many of the students projects are likely to include calling government agencies, polling community members' opinions, presenting information to the student goernment, parent-teacher organization or city council, or many other communication skills. If so, it may be important for students to practice these skills in class. Help them design role plays tailored to their needs.

Activity Procedures - Part D: Working on the Issue/Conducting the Project

a, *Create forums for encouragement and feedback.* Hold periodic sessions for progress reports. Encourage students to ask constructive questions or suggest strategies to each other. They may need ideas about whom to call for certain information, they may need to revise their plan and tasks, or they may have discovered infomation useful to another group.

b. *Announce important findings.* As students discover interesting data, get results from an experiment, or get a response from the governor, have them post it on a bulletin board or allow brief announcements to be made daily.

c. *Document their projects.* Have students document their work through photographs, videos, articles, etc. They should save questionnaires or other items created to build a portfolio of their projects.

d. *Encourage self-reflection.* Have students keep a journal throughout their projects. Teachers and students are likely to gain insight into personal challenges and progress through frequent journal entries.

e. *Keep in mind the importance of interviews and speakers.* Students can talk to town officials and to members of community, civic, and nonprofit organizations. They can talk to people representing different perspectives on the causes of and solutions for the problem. Talking with others in the community gives students a sense of what people are doing about the problem they are working on. They see people grappling with the problem and trying to solve it. The community becomes an integral part of the education process. Additional benefits include developing interviewing skills, being patient and persistent, speaking clearly, and listening carefully.

Students can conduct interviews with knowledgeable people about the problem. If the project is a class project, a speaker could be asked to come to class. Before an interview or a speaker presentation, have students prepare by thinking of questions to ask. It might be valuable to brainstorm some good interview questions and to role-play interviewing with students if they are not familiar with the process.

f. *Keep the projects focused.* To students, research and action can feel nebulous and overwhelming. The process is filled with frustrations, road blocks, dead ends, and wrong turns. The amount of information can seem endless. For these reasons, keeping the projects focused will be helpful.

Activity Procedures - Part E: Final Presentations

For Oral Reports

If students are working on independent group projects, the groups can report their findings and actions to the class when the projects are finished. This is a way of affirming their work and educating others about a specific issue related to trash disposal. Other students may want to participate in the action of a small group once it has been explained. Encourage students to include photographs, demonstrations, or other items that add evidence to their presentations.

For Written Reports

A finished paper is one way to evaluate students' work on this project and to help students practice writing skills. In addition, it asks students to organize and reflect on what they have learned. Papers may be written by each individual student or in small groups.

You may want students to write formal research papers that document findings and record information, or you may ask for a more informal piece of writing that explains the process of the research and action and provides a self-evaluation of the group's progress. Students can also simply respond to specific questions about their work or incorporate their answers into a unified piece of writing.

Some Suggested Structures for Presentaions

1. What made you choose this issue?
   How did you get information about the problem?
   Summarize what you learned about the problem.
   How did you decide what action to take?
   Describe what you did.
   What do you hope will result from your action?
   What did you learn from this experience as an activist?

2. What were the goals of your project?
   How did you organize your project?
   How did you publicize your project?
   Were you successful? Why or why not ? How do you know?
   What obstacles did you encounter? How did you overcome them?
   What did you learn from this experience as an activist ?

Publicizing the Projects

Students can also write up these projects as news reports or letters to the local school or local newspaper, explaining their work to others. This is one way for students to help educate the community about what they are learning and doing in school. Students might also write news letters to send home to parents or give to other students in the school.

A RANGE OF ACTIONS AND PROJECTS

Here is a list of possible projects and actions related to the issue of trash. Many options are open to any individual or group of people who want to make a difference on a particular issue. The following list does not include all of them by any means, but it is a starting place for ideas.

Inform others about the issues.
- Design a poster that will raise awareness about your issue.
- Write or perform a skit or dance to dramatize the importance of the issue.
- Set up a bulletin board about trash disposal issues in the school.
- Write and illustrate a book for younger children about the trash problem.
- Hold an assembly or public meeting to inform people about your issue.
- Create sculptures or songs to express your sentiments and draw attention to the issue.
- Make a videotape or slide show about the problem you are addressing.
- Publish and distribute a news sheet on the issue to raise awareness.
- Plan a day of education for your school or P.T.O. involving seminars or speakers. There are many possible subjects: household hazardous wastes, tips for reducing waste, the incinerator vs. landfill controversy, etc.

Write letters to those who influence or are effected by trash-related issues.
- Write to companies to commend their work on source reduction or recycling.
- Write to companies to encourage them to reduce waste.
- Write to local newspapers about local trash issues.
- Write to politicians about enacting laws on trash disposal.
- Write to young people in a town that has a hazardous waste site.

Start a recycling program in your school or community.
- Organize a white paper or aluminum can recycling center.
- Run a clothing, book, or toy exchange.
- Run a battery collection for recycling.

Volunteer to work for an organization that deals with your issue.

Do a "congress watch" to see what actions your representatives or other government officials have taken on your issue, and publicize the results.

Boycott a product, service, or business to express dissatisfaction and non-cooperation with its contribution to the trash problem.

Recruit materials needed to build composting structures. Announce to the school community that you are available with advice and materials to help families start composting at home.

HOW TO GET THINGS DONE

Organizing is the art of making something happen. Making a difference in the world is a challenge and a risk. You start off with an idea about the way you would like things to be, and then you take the risk of trying to bring that idea into reality. Organizing is also a means of creative expression. You add something new to the world to make it a better place.

Like any other worthwhile activity, organizing takes time, persistence, and practice. It also takes courage and, most of all, compassion for those around you. To help you be a better organizer, think about the following organizing principles:

The Dialogue Principle: When you are trying to create change, it is best to understand and empathize with those you are asking to change rather than to antagonize them. While an organizer takes a stand on an issue, he or she always needs to be able to say, "I may be wrong," be willing to listen, and, when appropriate, change his or her position. Change is difficult for most people.

People usually have good reasons for doing things the way they do, even if you find it hard to agree with those reasons. While you might think that a practice or policy is unfair or unjust and feel angry about it, the challenge of a good organizer is to be able to listen to and understand other people's perspectives and to express the problem in terms that they can appreciate and share. The best way to do this is to talk to people in a way that makes your concerns and interests clear and asks for their help and support in solving the problem. Instead of telling people what you want them to do, ask them to work with you on solving the problem you both share. For example, if the problem is that businesses are not buying recycled products, rather than going to local businesses and insisting that they buy such products, explain the problem to them and ask for their advice and help in solving it. This approach makes people feel appreciated, involved, and important. It demands from the organizer humility, patience, and a willingness to listen and ask for help.

The Support Principle: Organizing is hard work. It is usually more effective for a group of people to work together to make a difference than for someone to work alone. But working together takes talent and effort. A good organizer builds support within the group and also in the greater community.

When you are working together in a group, it is easier to think about the tasks that need to be accomplished rather than the people accomplishing those tasks. What makes people feel good is being appreciated. Make sure that you take the time to acknowledge people's efforts. If someone takes on a job that no one really wants to do, thank that person for being willing to do it. When someone accomplishes a task, have the group take the time to notice and thank that person. People do their best work when they know it matters to others and when they will be noticed and appreciated.

It also helps to build support outside your group. Ask other people for advice and help. Consult with those people who have decision-making influence and responsibility so that they feel included. Involving people in solving the problem with you often enlists their support for your efforts later on. Above all, express your appreciation for their help; people remember when they receive thank you notes.

The Leadership Principle: The most effective leader is the one who can work with a group in such a way that when the group has finished its work, the members feel they have done it themselves. Often leaders are seen as directors and authority figures. In organizing, the good leader is actually a facilitator rather than a director. This person shares leadership with others rather than controlling and dictating to the group. Instead of ordering and criticizing, this person listens to others' opinions and takes them into account, encourages people, notices their work, and appreciates them. This person helps people work together toward their common goal.

The greatest challenge for this leader is to keep the group on track and working effectively. This is accomplished by keeping the discussions focused, helping the group develop a schedule, and asking people how they are doing on their individual tasks. This leadership style is based on a desire for the group members to work well together, enjoy each other, and get the job done, rather than on a desire for power and recognition. Each and every one of us possesses the ability to be this type of leader.

The Principle of Appropriate Action: It is important to choose actions carefully. If you are doing something to benefit a specific group of people, first make sure that what you are doing is what they want. Second, make sure that your timing is appropriate. The right action at the wrong time never works very well -- a bit like asking a favor from someone who is angry with you. Third, be realistic about what you can accomplish. Use your resources and time effectively. Don't try to do everything. In many cases, your efforts will be one part of a larger effort, so find the piece that best fits your group's talents and interests. For example, if you wanted to change the amount of hazardous waste produced throughout your city or town, you would probably want to find out what specific action you could best contribute rather than taking on the whole issue. Finally, as a part of your project, inform the community about your efforts through the newspaper, TV, and radio. This will not only bring you additional support but will also show others that people can make a difference.

The Persistence Principle: Organizing is a challenge. Often it means doing things you are not used to or may not feel skilled at, such as talking to people in authority, analyzing statistics and information, or taking a stand. Trying new things can feel difficult and risky. A good organizer knows that he or she will make mistakes and learn from them, but what counts is persistence. The satisfaction of organizing comes from taking on a challenge and meeting it. Organizers also need to be flexible; as circumstances and plans change, you need to be able to bend, adjust, and learn as you go.

The Means-Are-the-Ends Principle: What you create will reflect the way you go about creating it. If you want to create a world in which people respect each other, then you must respect each other while you go about creating it. If you intimidate, antagonize, and coerce, then you may get what you want, but you will also get resentment, anger, and another battle later on. While it is appropriate to confront injustice, it is equally important to confront it justly. As a society, we progress only as fast as all of us progress; therefore, help people understand and join in your efforts.

GLOSSARY

Actinomycete - a lower organism similar to fungi and molds.

Activist - a person who energetically works for a cause; someone who attempts to raise awareness of an issue.

Aerate - to mix with air.

Aerobic - requiring or involving oxygen.

Aquifer - an underground geologic formation in which the cracks in rock, sand, soil, or gravel are filled with water.

Aseptic - a type of packaging which has 3 materials - aluminum, plastic, and paper - sealed together to keep out bacteria, light and oxygen.

Bacteria - single-celled organisms, some of which can cause disease. They can also break down solid waste.

Biodegradable - able to be broken down by microorganisms into simple, stable compounds.

Boycott - an organized effort to avoid supporting a person or an institution, done in the hopes that such an effort will pressure the person or institution to change a policy, decision, product, or other stance.

Carcinogenic - any substance that increases the risk of developing cancer.

Caustic - capable of burning or corroding living tissue by chemical action.

Cellulose fiber - a strand of a complex carbohydrate that forms the major part of the cell wall of most plants.

Classism - discrimination on the basis of socio-economic class which is systematically reinforced; the ideology which supports one group over another group on the basis of socio-economic class.

Clearcutting - the cutting down of all the trees in a given area.

Compost - a mixture of decayed organic matter used for fertilizer or soil conditioner.

Conserve - to protect from loss, to avoid wasteful or destructive use of.

Corrosive - (corrode (n)) - to eat away gradually, such as to rust.

Decay - to break down over time; when referring to organic matter, means the same as decompose.

Decompose - to separate or break down organic material into simpler molecules.

Decomposers - organisms that break down organic material.

Degradable - able to be broken down by some means, such as by microorganisms, sunlight, or heat.

Demographics - statistical information related to population, such as race, gender, age, income, education, profession, and so on.

Discrimination - the act or practice of according differential treatment to one group of people over another based on group identity, including class, race, religion, gender, sexual orientation, language, etc.

Emissions - substances discharged into the air as a result of burning.

Environmental racism - the term developed recently to identify the societal conditions which cause disproportionate effects of environmental hazards on people of color, and disproportionate exclusion of people of color in decision-making on environmental policies.

Evaporation - the process by which water is heated by the sun's energy and becomes vapor in the atmosphere.

Finite - having definite limits.

Flammable - easily set on fire.

Fly ash - the small particles of ash and soot that are collected by anti-pollution control devices in an incinerator or trash-to-energy facility.

Food Web - the flow of food energy from the roots of plants, which are then consumed by various levels of animals, to microorganisms which break down plant and animal wastes into nutrients which new plants take in through their roots.

Fungus (*pl.* fungi) - a many-celled or single-celled plant that is a saprophyte (lives on dead or dying materials) or parasite (lives on or in another organism) and lacks chlorophyll. This group includes molds, rusts, mildews, smuts, and mushrooms.

Garbage - spoiled or waste food that is thrown away; generally defined as wet food waste and excluding dry material (trash). This term is often used interchangeably with the word "trash."

Groundwater - water from beneath the earth's surface, between saturated soil and rock, that supplies wells and springs.

Groundwater contamination - the pollution of groundwater.

Hazardous - posing a danger to health and safety.

Hazardous wastes - wastes that, because of their chemical make up, can harm the environment.

Herbicide - a chemical for killing plants.

Humus - decayed plant or animal matter.

Incineration - the process of burning to ashes.

Incinerator - a facility where solid waste is burned.

Infinite - existing without boundaries or limits.

Inorganic - substances not made from plants, animals, or from carbon compounds (e.g., glass and metals).

Integrated waste management - the combined use of a variety of waste management practices to safely and effectively handle the municipal solid waste stream with the least amount of negative effects to human health and the environment. These practices include source reduction, recycling of materials, resource recovery, and landfilling.

Irritant - any substance that causes irritation, usually associated with redness and/or pain.

Landfill - a place where solid waste is disposed of by spreading it in thin layers and covering it with soil.

Leach - to percolate through soil or solid waste, carrying with it dissolved or suspended materials.

Leachate - a liquid that leaches through solid substance. It may carry hazardous wastes which can contaminate ground or surface water.

Lignin - the substance that forms the main part of wood in woody stems.

Microorganism - organisms too small to be seen with the naked eye. Microorganisms help break down compost.

Millbroke - paper scraps left from the process of making paper at paper mills. Millbroke has been recycled into other batches of paper throughout paper-making history. It is not post-consumer waste paper collected in paper recycling programs.

Monoculture - a habitat with only one variety of tree or other plant; it severely limits the number and variety of animal species which can live there.

Natural resources - valuable, naturally occurring materials such as timber, fresh water, and mineral deposits.

Non-biodegradable - not able to be broken down by organisms in a reasonable period of time.

Non-renewable resources - natural resources that take a long time to form, or are rapidly depleted, and as such are considered finite in amount. Examples: petroleum, coal, natural gas, and copper.

Nutrient - decayed plant or animal matter which enriches the soil.

Nutrient Cycle - continuous recycling of nutrients as they go from the soil to plants, then to animals, and back to the soil as decomposed matter and animal wastes.

Organic - a part of, or made from, any life form.

Oxygen Cycle - the continuous recycling of oxygen as it goes from the atmosphere to animals, then to plants, and eventually back to the atmosphere.

Packaging - the wrappings, containers, or sealing used to protect, identify, and advertise a product.

PCB's - polychorinated biphenyls, a chemical formerly used in electrical transformers, insulation for electric cables, and production of electric condensers.

Pesticide - a chemical for killing unwanted animals, such as insects and rodents.

Photodegradable - degradable by sunlight.

Photosynthesis - the process by which plants make "food" (carbohydrates) using carbon dioxide, water, and sunlight.

Post-consumer waste - trash as the result of consumer use.

Precipitation - water in the earth's atmosphere falling from clouds in the form of rain, hail, sleet, or snow.

Prejudice - an attitude, feeling, or opinion formed without sufficient knowledge or reason. It is a judgment made prematurely, a predisposition.

Primary consumer - a plant-eating organism.

Protozoa - the simplest form of animal organisms. They are single-celled and microscopic but more complex in their activities than most bacteria.

Racism - discrimination on the basis of race which is systematically reinforced; the ideology which supports one group over another group on the basis of race.

Reclamation - the restoration to usefulness or productivity of materials found in the waste stream.

Recyclable - able to be recycled. Includes almost all natural and human made materials like glass, aluminum, paper, plastic, oil, tin.

Recycle - to put through a cycle again, to collect and reprocess materials for reuse in either the same form or a different form.

Recycling - the collection and processing of materials for reuse either in the same form or as part of a different product.

Renewable resources - natural resources derived from an endless source (sun, wind, wood, fish) or able to be replaced by new growth.

Resource - valuable material; a source of wealth.

Resource recovery - a process by which energy is produced by burning solid waste.

Reusable - reuse - to extend the life of an item by repairing or modifying it or by creating new uses for it.

Reusing - extending the life of an item by creating new uses for it.

Runoff - rain or melted snow that flows over the land.

Secondary consumer - an animal eating organism that eats the primary consumers.

Selective cutting - the cutting down of only certain trees in a given area based on a criteria such as size, age, or location within the area.

Sludge - a muddy sediment left after sewage has been processed.

Slurry - a mixture of blended paper and water used in making paper.

Solid waste - any human- or industry-generated waste excluding that from toilet systems (before it has been processed) and drinking/washing water.

Solid waste management - the system a region (e.g. city, county, state) or company has for treating and disposing of its waste.

Source reduction - the deliberate reduction of waste by making choices in production, consumption and disposal of items.

Stereotype - a generalization about the characteristics of a group of people. They are assumed to describe everyone in that group without distinguishing any individual's unique characteristics.

Tertiary consumer - an animal-eating organism that eats the secondary consumers.

Tipping fee - the charge to deposit waste in a landfill.

Toxic - poisonous.

Transpiration - the process by which water vapor is released by plants into the atmosphere.

Trash - material considered worthless, unnecessary, or offensive that is usually thrown away; generally defined as dry waste material, excluding food waste (garbage) and ash. This term is often used interchangeably with the word "garbage."

Trash-to-energy - see "resource recovery."

Unlimited - having no limits, bounds or qualifications.

Virgin Paper - paper that has never been recycled, made directly from tree pulp.

Waste incinerator - see "incinerator."

Waste-to-energy - see "resource recovery."

Watershed - a region or area surrounding a body of water that drains ultimately into the body of water.

Water table - the surface level of groundwater. It varies in level, for example, it is highest during the spring after snows melt and rains fall; it is lowest after dry summers.

RESOURCE DIRECTORY

JOURNALS

E. *The Environmental Magazine.* P.O. Box 6667, Syracuse, NY 13217-7934.
A non-technical magazine covering current environmental issues.

Garbage Magazine. 435 Ninth St. Brooklyn, NY 11215
A magazine written for adults on a variety of topics related to waste management.; written in non-technical language.

BOOKS

Berthold-Bond, Annie. *Clean and Green: The Complete Guide to Nontoxic and Environmentally Safe Housekeeping.* Ceres Press, P.O. Box 87, Woodstock, NY 12498. 1990.
An excellent guide, with recipes and lists of places to get hard-to-find items.

Cohen, Gary and John O'Connor, eds. *Fighting Toxics: A Manual for Protecting Your Family, Community, and Workplace.* Island Press, Suite 300, 1718 Connecticut Ave. NW, Washington, DC 20009. 1990.
This book gives specific information on how to organize your community, influence corporate business practices, and obtain information on the threats in your area.

Council on Economic Priorities. *Shopping for a Better World.* 30 Irving Place, New York, NY 10003. (800) .729-4237. 1992.
This pocket-size guide lets consumers know where companies stand on issues ranging from animal testing to the efficient use of energy. Also available is the new *Students for a Better World* (1993) which includes companies young people are particularly likely to buy from.

Elkington, John. *The Green Consumer.* Viking Penguin, 40 West 23rd St., New York, NY 10010. (212) 337-5200. 1990.
This book provides a shopping guide to a wide range of environmentally safe or low-impact consumer goods.

Minnich, Jerry, and Marjorie. Hunt. *The Rodale Guide to Composting.* Rodale Press, Emmaus, PA. 1979.
This comprehensive guide will answer all your questions about composting.

Natural Resources Defense Council. *50 Simple Things You Can Do To Save the Earth.* The Earth Works Group, Earthworks Press, Box 25, 1400 Shattuck Ave Berkeley, CA 94709 (415) 841-5866. 1989.
A simple "hands-on" guide on both big and small things you can do to help the environment. Also available: *50 Simple Things Kids Can Do To Save the Earth.*

U.S. Department of Commerce, Bureau of the Census. *1980 Census of Population. General Population Statistics: United States Summary,* issued May 1983. *General Social and Economic Characteristics: United States Summary,* issued December 1983. These volumes and many others are available from regional offices of the Census Bureau and Government Printing Office bookstores, as well as reference libraries. They include vast amounts of demographic data on the U.S. population. Data from the 1990 census is now available, but with less detail then the 1980 census.

BOOKLETS ON SPECIFIC TOPICS

Concern, Inc., 1794 Columbia Road NW, Washington DC, 20009. (202) 328-8160. Concern has booklets that are good for background information. Topics include: household waste, waste, farmland, drinking water and groundwater.

SOURCES OF RECYCLED PAPER AND OTHER RECYCLED PRODUCTS

Atlantic Recycled Paper Co., P.O. Box 11021, Baltimore, MD 21212. (800) 323-2811.

Conservatree Paper Co. 10 Lombard St. Suite 250, San Francisco, CA 94111, (800) 522-9200.

Cross Pointe 18 E 41st St., Suite 1801, New York, NY 10017. (212) 213-1536.

Earth Care Paper Co., Box 3335, Dept. 42, Madison, WI 53704. (608) 223-4000.

Fay Paper, 59 Davis Ave., Norwood, MA 02062. (800) 532-9573.

Recycled Paper Co., 185 Corey Road, Boston, MA 02146-9901. (617) 277-9901.

Seventh Generation, 10 Farrell St., Dept. MO4012, South Burlington, VT 05403. (800) 456-1177.

Very Important Paper, Box 90, Cuttingsville, VT 05738. (802) 492-3304.

HOUSEHOLD CLEANING PRODUCTS

Ecover (brand) - available at natural and select food stores; distributed by Mercantile Food Company, Georgetown, CT 06829.

NATIONAL ENVIRONMENTAL ORGANIZATIONS

Citizens Clearinghouse for Hazardous Waste, P.O. Box 926, Arlington, VA 02216.

Citizens for a Better Environment, 33 East Congress Parkway, Suite 523, Chicago, IL 60605.

Clean Water Action, 317 Pennsylvania Ave. SE, Washington, DC 20003.

Earth First!, P.O. Box 5176, Missoula, MT 59806.

Environmental Action, 1525 New Hampshire Ave. NW, Washington, DC 20036.

Environmental Defense Fund, 257 Park Ave. S., New York, NY 10010.

Friends of the Earth/Environmental Policy Institute, 530 7th Ave. NW, Washington, DC 20009.

Greenpeace, 1432 U St. NW, Washington, DC 20009.

National Toxics Campaign, 1168 Commonwealth Ave., Brighton, MA 02135. (617) 232-4014.

Natural Resources Defense Council, 122 East 42nd St., New York, NY 10168.

Sierra Club, 730 Polk St., San Francisco, CA 94109.

Bibliography

Books

Bullard, Robert D., <u>Confronting Environmental Racism</u>, South End Press, Boston, MA, 1993

Bullard, Robert D., <u>Dumping in Dixie: Race, Class, and Environmental Quality</u>, Westview Press, Boulder, CO, 1990

Martin, Debra L. and Grace Gershung, Eds., <u>The Rodale Book of Composting: Easy Methods for Every Gardener</u>, Rodale Press, Emmaus, PA, 1992

Rosenberg, Charles E., <u>The Cholera Years</u>, University of Chicago Press, 1962

Suess, Dr., <u>The Lorax</u>, Random House, New York, 1971

Resource/Study Guides

Bell, Carol O. and Martha J. Schwartz, <u>Oscar Options - Book II</u>, Ocean State Cleanup and Recycling, Department of Environmental Management, Providence, RI, 1987

Boston Area Educators for Social Responsibility, <u>Making History: A Social Studies Curriculum in the Participation Series</u>, Cambridge, MA, 1984

Department of Environmental Protection, <u>The Solid Waste Management Resource Guide for Massachusetts Schools</u>, The Commonwealth of Massachusetts Executive Office of Environmental Affairs, Boston, MA, 1990

Hallowell, Ann, et al., <u>Recycling Study Guide</u>, Education Programs, Bureau of Information and Education, Wisconsin Department of Natural Resources, 1989

Pedicord, Susan, <u>AVR Teacher's Resource Guide for Solid Waste and Recycling Education</u>, Association of Vermont Recyclers, Montepelier, VT, 1984

Schneidewind, Nancy and Ellen Davidson, <u>Open Minds to Equality</u>, Allyn and Bacon, Boston, MA, 1983

Water Pollution Control Federation, <u>Household Hazardous Wastes: What You Should and Shouldn't Do</u>, Alexandria, VA, 1987

Newspapers and Magazines

Arias, Ron and Liz McNeil, "A boy sides with Dr. Seuss's Lorax and puts a town at loggerheads", *People Magazine*, Vol. 32, Oct. 23, 1989, pg. 67-68

Bellafante, Ginia and Janet Marinelli, "The Good, the Bad, and the Ugly", *Garbage Magazine*, May/June, 1991

Boston Globe, "Indian reservations are targeted for nation's waste", June 28, 1991

Boston Globe, "School decides to monitor pests rather than clobber them", June 23, 1991

Elson, Joan, "Dumping on the Poor", *Time*, August 13, 1990, pg. 46

Environmental Professional, V 11, 1989

Goodman, Billy, "Fill 'er Up", *Science World*, May 6, 1988

Ideas: A compendium of 3M Success Stories, "Wastestopper: Pumice on Copper", Environmental Engineering and Pollution Control Dept./3M, Saint Paul, MN

Kolata, Gina, "Tree Yields a Cancer Treatment, But Ecological Cost May Be High", *New York Times*, May 13, 1992

New York Times, January 26, 1989

New York Times, September 8, 1991

Newsweek, November 27, 1989

One Person's Impact, "Saying 'No': Boycotts and consumer action", June/July 1990

People Extra, "Earthen Renewal Begins at Home", Spring 1990
Science World, "Bag It", April 20, 1990
Time, "Garbage, Garbage Everywhere", September 5, 1988
Winslow Mail, "Navajos Vote Against Toxic Waste Facility", Winslow, AZ, May 4, 1990

Government Documents

Department of Environmental Protection, Outline for a Home Composting Presentation, Boston, MA, 1993

Massachusetts Water Resources Authority and The Museum of Science, Boston, Water Watchers, Boston, MA, 1991

U.S. GAO, Siting at Hazardous Waste Landfills and Their Correlation With Racial and Economic Status Surrounding Communities, 1983

Videos

Fire in Their Hearts, available from Boston Area ESR, (617) 492-8820
The Lorax, available from any EPA and some video stores

Other

Cross Cultural Consultation. Phone conversation, Randolph, MA 1993
PC Connection Catalog, Fall 1990, PC Connection, Marlow, NH
United Church of Christ Commission on Racial Justice, Toxic Waste and Race in the United States: A National Report on the Racial and Social Economic Characteristics of Communities with Hazardous Waste Sites, Cleveland, OH, 1987

SECTION 313 EPA REGIONAL CONTACTS

Region 1

Pesticides & Toxics Branch
USEPA Region 1 (APT2311)
JFK Federal Building
Boston, MA 02203
(617) 565-4502

Connecticut, Massachusetts, Maine, New Hampshire, Rhode Island, Vermont

Region 2

Pesticides & Toxics Branch
USEPA Region 2 (MS240)
Woodbridge Avenue, Building 209
Edison, NJ 08837-3679
(201) 906-6890

New Jersey, New York, Puerto Rico, Virgin Islands

Region 3

Toxics & Pesticides Branch
USEPA Region 3 (3HW42)
841 Chestnut Street
Philadelphia, PA 19107
(215) 597-1260

Delaware, Maryland, Pennsylvania, Virginia, West Virginia, District of Columbia

Region 4

Pesticides & Toxics Branch
USEPA Region 4
345 Courtland Street
Atlanta, GA 30365
(404) 347-1033

Alabama, Florida, Georgia, Kentucky, Mississippi, North Carolina, South Carolina, Tennessee

Region 5

Pesticides & Toxic Substances Branch
USEPA Region 5 (5SPT-7)
230 South Dearborn Street
Chicago, IL 60604
(312) 353-5907

Illinois, Indiana, Michigan, Minnesota, Ohio, Wisconsin

Region 6

Pesticides & Toxic Substances Branch
USEPA Region 6 (6TPT)
1445 Ross Avenue
Dallas, TX 75202-2733
(214) 655-7244

Arkansas, Louisiana, New Mexico, Oklahoma, Texas

Region 7

Office of Congressional and Intergovernmental Liaison
USEPA Region 7 (CIGL)
726 Minnesota Avenue
Kansas City, KS 66101
(913) 551-7005

Iowa, Kansas, Missouri, Nebraska

Region 8

Toxic Substances Branch
USEPA Region 8 (8AT-TS)
999 18th Street
Denver, CO 80202-2405
(303) 293-1730

Colorado, Montana, North Dakota, South Dakota, Utah, Wyoming

Region 9

Pesticides & Toxics Branch
USEPA Region 9 (A-4-3)
75 Hawthorne Street
San Francisco, CA 94105
(415) 556-5387

Arizona, California, Hawaii, Nevada, American Samoa, Guam, Commonwealth of the Northern Mariana Islands

Region 10

Pesticides & Toxic Substances Branch
USEPA Region 10 (AT083)
1200 Sixth Avenue
Seattle, WA 98101
(206) 553-4016

Alaska, Idaho, Oregon, Washington